W9-BKW-583

HISTORIANS AND
EIGHTEENTH-CENTURY
EUROPE
1715–1789

HISTORIANS AND EIGHTEENTH-CENTURY EUROPE
1715-1789

M. S. ANDERSON

CLARENDON PRESS · OXFORD
1979

Oxford University Press, Walton Street, Oxford OX2 6DP

OXFORD LONDON GLASGOW
NEW YORK TORONTO MELBOURNE WELLINGTON
IBADAN NAIROBI DAR ES SALAAM CAPE TOWN
KUALA LUMPUR SINGAPORE JAKARTA HONG KONG TOKYO
DELHI BOMBAY CALCUTTA MADRAS KARACHI

© *Oxford University Press 1979*

*All rights reserved. No part of this publication may be reproduced,
stored in a retrieval system, or transmitted, in any form or by any means,
electronic, mechanical, photocopying, recording, or otherwise, without
the prior permission of Oxford University Press*

British Library Cataloguing in Publication Data

Anderson, Matthew Smith
   Historians and eighteenth-century Europe, 1715–1789.
   1. Europe – History – 18th century – Historiography
   I. Title
   940.2′53′072   D288   78–40319

   ISBN 0–19–822548–2

*Printed in Great Britain by
Billing & Sons Limited, Guildford, London and Worcester*

PREFACE

It will be clear at once to the reader that this book is not a conventional history of the period which it covers. It is instead something perhaps more ambitious and certainly in many ways more difficult to write; a history of the way in which thinking about the main aspects of the history of eighteenth-century Europe has developed over the last two centuries or more. The sheer vastness of the materials available makes the writing of such a book a difficult task, however rewarding. I have therefore concentrated on a number of central issues which have attracted sustained interest on the part of historians and around which historical controversy has tended to centre: in each of these I have tried to show how historiography has developed and views have altered with the progress of time. The book ends with a Conclusion which attempts, very briefly, to outline the present position of historical writing about this period.

Work on this book has given me a clearer understanding of two unsurprising but none the less important facts. The first is the extent to which history is a subject which grows by slow accretions, through the progressive building-up by sustained work of a more complex and accurate body of knowledge and a more wide-ranging and sophisticated system of ideas. Few great works of history are totally revolutionary, in the sense of overturning completely and without warning existing views of any major aspect of the past. In the period covered by this book even a Tocqueville, a Namier, or a Labrousse has not been able to accomplish quite such a transformation. Very often, once the initial shock of a radical new approach has been tempered, it is found that it can be absorbed without too much difficulty into an existing intellectual structure. That structure is modified, perhaps even fundamentally changed, but seldom completely abandoned. The second conclusion which has been forced on me, and which in a sense follows from the first, is an optimistic one. It is the extent to which, as a result of this process of increasing knowledge and developing ideas, our understanding of eighteenth-century Europe is now far superior to anything within the reach of past generations. It is not merely that our knowledge of the facts, in a narrow sense, has been immensely expanded by decades of patient library-haunting and archive-searching. More important, our view of what facts are worth knowing and can be known is far wider and far more realistic than ever in the past. Historians during the twentieth century have emancipated themselves from a restrictingly political view of their subject. Not merely is the economic history of eighteenth-century Europe now being explored with a thoroughness and pene-tration not previously possible; not merely is European society of that age being studied with the aid of a whole range of new tools and concepts; but the mind of the age in its broadest sense, the way in which ordinary people thought and felt, is now, at least in some parts of the continent, the object of scholarly analysis with very exciting results. History as an academic discipline advances less dramatically than the physical sciences, with fewer 'great discoveries' and less obvious discontinuities. But it does advance none the less; and the advances are solid and permanent ones.

I must end the preface to this book by making clear my debt to the libraries without access to whose resources it could not have been written. The British Library, Reference Division (formerly the British Museum Library) and the London Library have both been of great help. Above all I have made heavy demands on the British Library of Political and Economic Science, the library of the London School of Economics. It has been for many years a privilege, as well as a great intellectual advantage, to serve an institution which possesses the finest specialized collection in Europe, perhaps in the world, of material on the social sciences. I have never been more conscious of that fact than while working on this book.

London School of Economics   M. S. ANDERSON

CONTENTS

CHAPTER I

THE OLD REGIME IN FRANCE

IN EIGHTEENTH-CENTURY Europe France was not merely the greatest
of states but the greatest of all influences on intellectual and artistic
life, the centre from which more than any other the mind and
emotions of the Continent were shaped. In the physical sciences
and in political and social speculation she was pre-eminent, even
though her pre-eminence was based on foundations laid on the other
side of the Channel by Newton and Locke. Until well into the second
half of the century Louis XIV appeared, to most Europeans, the
model of what monarchy might be and indeed ought to be. In painting, in architecture, in all the minor arts which did so much to
determine the quality of daily life for the ruling minority, above all
in literature, the position of France was dominant, often overwhelming. No European state has made itself so completely and in
so many aspects of life the arbiter of an epoch as France during the
three or four generations before the Revolution. The greatness of her
position has been reflected in a body of writing on her history, her
strengths and weaknesses, her problems and triumphs during these
generations, which in extent and quality can hardly be approached,
far less paralleled, in the case of any other European state of the
period. In any study of what historians have thought and believed
about eighteenth-century Europe, of how their knowledge has
deepened, their ideas have broadened, their preconceptions have
altered, she must play a leading role.

Writers on the life of eighteenth-century France have from its
beginnings had to resist a deeply seductive temptation. The old
regime collapsed at the end of the century in an unprecedented
cataclysm and by so doing unleashed a wave of political and social
change whose ripples could be felt in the most distant and backward
parts of the Continent and even outside it. It has therefore been only
too easy for historians to see the eighteenth century in France as a
mere preparation for the Revolution, simply as the story of a movement, inevitable and irresistible, towards the breakdown at the end
of the 1780s. Not until our own time have they freed themselves,

in so far as they have ever done so completely, from assumptions of this kind, assumptions all the more pervasive and misleading since they were seldom if ever explicitly stated. A full recognition of the last generations of the old regime as an age of French history complete in itself and with its own character and problems has come late and slowly. For a century or more after 1789 it was too easy and too tempting to see this period as a mere preparation for the revolutionary and Napoleonic drama, as a kind of interlude between the two completely different high points represented by Louis XIV and by the Jacobins.

EARLY WORKS: FROM THE EIGHTEENTH CENTURY TO THE JULY MONARCHY

Certainly the relatively few writers of the reigns of Louis XV and Louis XVI who attempted to tell the story of their own age cannot be said to have added a great deal to the historian's stock of ideas or insights. Attempts at systematic writing of the history of eighteenth-century France were made, for example, by C. J. F. Hénault in his *Abrégé chronologique de l'histoire de France* (Paris, 1744) and by A. E. N. Fantin des Odoards in his *Histoire de France depuis la mort de Louis XIV jusqu'à la paix de Versailles de 1783* (8 vols., Paris, 1789); but these were no more than annals of the most aridly factual and intellectually limited kind. Both show some grasp of the fact that politics, diplomacy, and war do not make up the whole of the history of France during this period;[1] but both are concerned overwhelmingly with the struggles in Paris and Versailles of different individuals and factions for political power, with foreign policy and military events. Both therefore see events entirely from the standpoint of the capital and the government; the diversity of provincial France, which in the eyes of many twentieth-century historians is both the greatest difficulty and the greatest fascination of the study of the old regime there, is completely lacking in these laborious and blinkered compilations.

Nor does the more specialized historical writing of the period go very far to meet the demands of the present-day scholar. C. Pineau-Duclos, in his *Mémoires secrets sur les règnes de Louis XIV et Louis XV* (Paris, 1791) could claim[2] to have used much unprinted material

[1] For example, the long summary list of French achievements in science, technology, intellectual life, colonial expansion, etc., given by Fantin des Odoards, i. 25–48.

[2] Preface, p. xvii.

in the form of diplomatic correspondence and memoirs (including those of Saint-Simon). Nevertheless he produced merely a narrative of an anecdotal kind, centred entirely on the court and Paris and with a heavy emphasis on the foreign policy of the period down to the formation of Fleury's ministry in 1726 (the author died in 1772 leaving the book unfinished). There was a good deal of writing in particular on John Law and his spectacular and forward-looking financial schemes during the Regency. But all of this is polemical, often violently so, and usually strongly hostile to Law.[1] Except as an index of widely-held and influential attitudes towards economic and above all financial affairs, its value to the historian of the old regime is limited. Contemporary biographies of French statesmen tend to be eulogistic, often highly so: Condorcet's *Vie de Monsieur Turgot* (London, 1786) and C. J. de Mayer's *Vie politique et privée de Charles Gravier, Comte de Vergennes* (Paris, 1789) are obvious examples of this. Neither is without merit. That by Condorcet is founded on personal knowledge of and admiration for Turgot, while Mayer made a real effort to base his work on documentary evidence, some of which he reproduced in his book. But no author before the Revolution attempted a study of the France in which he lived, or of any large aspect of its life, in terms which the historian of today finds intellectually satisfactory.

The coming of the Revolution, the Terror, above all the emigration, completely changed the situation. It now became an urgent preoccupation of many Frenchmen, particularly among the *émigrés*, to discover the reasons for the catastrophe which had overwhelmed France, the monarchy, and themselves. This inevitably meant evaluating the old regime, deciding how far its own weaknesses had been responsible for its downfall. Two diametrically opposed attitudes began to show themselves almost at once. On the one hand it was urged that the old regime had been not merely sound but beneficent and progressive. The Revolution was therefore the outcome essentially of the malice of a few discontented individuals, aided by fortuitous events and by the errors, not necessarily serious in themselves, of those in power. On the other hand it was argued that the disintegration of the old regime was the inevitable result of its own

---

[1] J. Paris du Verney, *Examen du livre intitulé reflexions politiques sur les finances et le commerce* (2 vols., Paris, 1740), Dupont de Nemours, *Du Commerce et de la compagnie des Indes* (Paris, 1769), and L. L. F. de Brancas, Duc, *Mémoire sur la compagnie des Indes* (Paris, 1769) are good examples.

defects and of new forces and a new atmosphere which had been
developing in France for at least several decades. The revolution
could be seen, in the most extreme form of this argument, as the
climax of processes of change extending over centuries.

The most detailed and convincing statement of the first of these
points of view came as early as 1795 when the moderate monarchist
émigré, G. Sénac de Meilhan, published in London his *Du gouverne-
ment, des mœurs, et des conditions en France avant la révolution*.
Whatever theorists might say, he argued, and however marked the
contrast with England (which as a political model he greatly admired)
France in the seventeenth and eighteenth centuries was not a
despotism. It saw during these centuries great economic progress
which could not have been achieved under despotic rule, and also
a wide-ranging freedom to attack authority visible in an increasing
profusion of criticism and satire. There was also a real division of
political powers within the country, since the *parlements* balanced
military influences and the intendants carefully scrutinized the
actions of the judiciary; and this division was a guarantee of liberty.[1]
The Third Estate had not been oppressed, humiliated, or deprived of
outlets for its talents. Nor were the different social classes rigidly
separated as had been claimed by some of the revolutionaries, for a
man of ability could always attain a position entitling him to respect.[2]
Even the venality of offices did on balance more good than harm,
since it provided a guarantee that the office-holder possessed some
education and wealth which minimized the dangers of his dependence
on the great and of outright corruption. The essential causes of the
Revolution could be reduced to three, all essentially superficial and
even fortuitous: the writings and conduct of Necker, which inflamed
the situation; the 'too easy good nature of the king'; and the actions
of the Assembly of Notables of 1787. The writings of the *philosophes*
and the development of French intellectual life had little to do with
the matter. The works of Voltaire, though they had shaken faith in
religion, had had little relevance to government and had been on the
whole favourable to monarchy as an institution.[3] Montesquieu had
been an apologist for monarchy, and for the nobility and *parlements*.
Rousseau's *Social Contract*, Sénac de Meilhan admitted, contained
ideas which were in conformity with those of some of the revolutionaries; but it had been little read or understood before 1789. Mably,
perhaps the most genuinely revolutionary of the writers of the eight-

[1] pp. 23–5.   [2] Chaps. vii–viii, *passim.*   [3] pp. 84, 100, 102.

eenth century, had enjoyed little popularity.[1] It was only after the Revolution had taken place that the works of Rousseau and Mably became significant as providing ideas which helped to sustain it. This almost contemptuous writing-down of the influence of the *philosophes* at a time when other commentators were widely exaggerating it gives the book in some ways a markedly modern ring. Necker, the author reiterated, had by his errors and ambitions been largely responsible for the fate which had overtaken France and the monarchy. In particular he had decided that the States-General should meet at Versailles; and this, because of the proximity and disastrous influence of Paris, had had a decisive influence on events. He had, in effect, decided that the Third Estate should become dominant because he hoped to profit personally from its dominance.[2] Other factors had contributed to the débâcle. The court under Louis XVI had lost much of its former glamour, and had thus weakened the belief that there were social distinctions which did not depend and ought not to depend merely on wealth. Turgot again, for all his great abilities, had introduced into government in France a dangerous element of ideology, 'l'esprit de secte'. The character of the French people, frivolous, changeable, at the mercy of fashion, inclined to push all ideas, once adopted, to extremes, must also bear some of the blame.[3] All this added up to a forcible assertion that the old regime was sound at heart and that the Revolution, the product of pride and irresponsibility, had no real justification and little root in French history or in the realities of the pre-revolutionary situation.

No other writer of the period developed this line of argument so systematically and effectively. One aspect of it in particular, however, was frequently repeated by opponents of the Revolution—the belief that the monarchy had been weakened, perhaps doomed, by the deficiencies of its ministers in the generation before 1789. Thus, the *émigré* A. F. de Bertrand-Moleville in his strongly royalist *Mémoires particuliers pour servir à l'histoire de la fin du règne de Louis XVI*

---

[1] pp. 128–9.

[2] p. 185. There is a long and highly unfavourable picture of Necker on pp. 172–95. In the hands of more imaginative and less scrupulous writers his role in the events of 1788–9 became proof of a plot by radical *émigrés* from Geneva, subsidized by Britain, to overthrow the French monarchy, e.g. J. L. Soulavie, *Historical and Political Memoirs of the Reign of Lewis XVI* (London, 1802), v. 280–3, and chap. xv, *passim*.

[3] Sénac de Meilhan, pp. 133, 141–2, 134–8.

(Paris, 1816) argued that all the reforms the French people needed in the later eighteenth century—the abolition of *lettres de cachet*, equitable taxation, equal rights of all citizens, recognition of the need for taxation to be agreed to by the States-General—could have been achieved peacefully if only Louis XVI had possessed better ministers. In particular Maurepas had, in the interests of his own position, stultified the natural good qualities of the king, encouraged his timidity and lack of self-confidence, limited his grasp of government business, and stifled his energies.[1] Some years later, under Charles X, a much more interesting writer, Alexandre de Lameth, who had himself played an important role in the early stages of the Revolution as a moderate reformer, developed, in the Introduction to his *Histoire de l'assemblée constituante* (Paris, 1828) a different and much more fruitful line of argument. This was that the old regime, essentially viable, had been destroyed in part by its own weakness and lack of decision (seen for example in its failure to intervene actively in the Dutch Republic in 1787 and thus begin a war which would have strengthened its position in France) and above all by the selfishness of the nobles and clergy, its natural and traditional allies. These groups in 1786–8 worked against the king and the government in the interests of their own position and power. When in autumn 1788 they realized how dangerous the situation had become and rallied to the support of the monarchy it was too late; by placing themselves and the remnants of feudalism in France under the protection of the crown they merely increased its difficulties. It was they who had begun the Revolution; if it were a crime they above all were guilty of its outbreak.[2] This was an analysis of events which, however sound and obvious it may seem today, aroused little response when it was first put forward. But like those of Sénac de Meilhan and a numerous throng of counter-revolutionary writers it assumed that the old regime had been destroyed by the selfishness and frivolity of identifiable groups and individuals, by decisions taken shortly before the outbreak of the Revolution, in a word by forces which it was neither necessary nor possible to trace very far back into the history of France.

Side by side with beliefs of this kind, however, existed others which postulated that the old regime had for long been undergoing change, above all in its intellectual climate, which had progressively

[1] i. 17–18, 25–30.
[2] i. Introduction, pp. lxxvii, xciii–ci.

undermined the entire structure. More and more it was argued by writers of this persuasion that the Enlightenment was to blame for what had happened in France. As early as 1792 the revolutionary Rabaut Saint-Étienne, in his *Almanach historique de la révolution française pour l'année 1792*, had presented the Revolution for the first time as the outcome of a long process of intellectual growth to which Bacon, Bayle, even Montaigne, had all contributed. It was thus in a sense the outcome of the entire previous history of France and not the mere caprice of an unstable people, as some foreigners liked to imagine.[1] By the first years of the nineteenth century this line of thought was widespread; in particular attacks on the *philosophes* for their arrogant rejection of traditional intellectual restraints, for their irreligion, and for the socially disintegrating effects of their doctrines were becoming commonplace. They 'limited the power of heaven itself' complained one writer, and adroitly set in motion the Third Estate which otherwise would have remained quiescent.[2] They were a conceited and intolerant sect, alleged another, which had failed to support Louis XV and Louis XVI while becoming 'the slave and pensioner of two foreign rulers' (George III and Joseph II are meant). They had thus undermined patriotism and replaced it by cosmopolitanism, while their internal feuds—e.g. those of Voltaire with Mably or Rousseau—had foreshadowed the later factional conflicts of the revolutionaries.[3] It was above all, complained another and more balanced commentator, the vagueness of the idea of liberty which they had implanted in France which was dangerous, since this lack of precise and limited objectives led easily to excess in efforts to realize the idea in practice.[4]

Prosper de Brugière de Barante, in his frequently reprinted *De la littérature française pendant le dix-huitième siècle* (Paris, 1809) summed up ably and with moderation many of these accusations. Precisely because the *philosophes* had sincerely wished to benefit mankind they had been intolerant and excessively self-confident;

[1] See the discussion and quotation in A. Aulard, *Études et leçons sur la révolution française*, 6th ser. (Paris, 1910), p. 50.

[2] C. F. Beaulieu, *Essais historiques sur les causes et les effets de la révolution de France* (6 vols., Paris, 1801–3), i. Introduction, xlvii, xxxiv.

[3] J. L. Soulavie, *Histoire de la décadence de la monarchie française* (Paris, 1803), iii. 108–9, 122; cf. his *Historical and Political Memoirs . . .*, vi. 89, where he complains of their having 'transformed the monarchical genius of France into a spirit of anarchy and rebellion'.

[4] F.-E. Toulongeon, *Histoire de France, depuis la révolution de 1789* (Paris, 1801), i. 18.

moreover, unlike earlier thinkers they had not been remote or unwordly but rather men of the world, rapid and superficial in judgement, contemptuous of erudition and the past, deeply influenced by the overweening vanity which was so visible in their quarrels and polemics.[1] But the change which they had produced in the climate of opinion in France had been of fundamental importance; for revolutions occurred when the slow march of ideas suddenly found itself out of step with existing institutions. Even before the end of the reign of Louis XIV political authority and religion had begun to lose their grip; scepticism was increasing, judgements of all kinds were becoming more facile but losing weight and solidity, and individuals were beginning to value their own ideas more than received ones. The Revolution was thus the outcome of a slow change of attitude in France and of a growing desire for political change there. Its roots, which were ultimately intellectual, went back for several generations. It was not the result of a specific crisis or of simple malice or miscalculation on the part of individuals. On the contrary, its leaders acted from the highest motives, believing that the changes they hoped for would be quite easy to accomplish. It was only when new men from the lower classes, eaten up by vanity and envy, rose to prominence that its character altered drastically for the worse.[2] This attribution of the fall of the old regime mainly or entirely to the march of intellectual change was an attitude which was to enjoy a long and in some ways productive life.[3]

In other ways also it was possible to see the Revolution as the outcome of the whole history of France from the end of the Middle Ages. Mme de Staël, the daughter of Necker, put forward a superficial view of this kind in her *Considérations sur les principaux événements de la révolution française* (London, 1818), in which she presented the Revolution simply as the outcome of a desire for

[1] pp. 111, 141–4, 107–8, 188.

[2] pp. 5–6, 47, 255–6.

[3] For later manifestations of it see below, pp. 24–6, 31–3. The most extreme and distorted form taken by this faith in the ability of ideas to determine events was of course the belief that the old regime had been brought down by a masonic (in early forms of the belief also Protestant) plot. The Abbé Barruel's famous *Mémoires pour servir à l'histoire du jacobinisme* (London, 1797), the best-known statement of this belief, was effective largely because it brought together a considerable body of suspicions of this kind already in circulation. The subject has recently been explored in detail in J. M. Roberts, 'The Origins of a Mythology: Freemasons, Protestants and the French Revolution', *Bulletin of the Institute of Historical Research*, xliv. No. 109 (May 1971). As it has only marginal relevance to the historiography of the old regime it is not discussed further here.

political liberty and for an effective constitution. The whole history of France had been in effect a series of efforts by the nation to assert its rights and by the nobility to maintain its privileges, while most of the kings had tried incessantly to make themselves absolute.[1] It was the efforts of Richelieu and Louis XIV to establish arbitrary power, the struggles of the privileged in the years before 1789 to weaken the monarchy, and the reaction of the nation to these developments, which explained the cataclysm of the last years of the eighteenth century. But this was a mechanical and superficial view of the situation: the book was weakened by Mme de Staël's marked desire to defend the memory of her father and by her dogmatism and preoccupation with constitutional considerations to the exclusion of all others. A much more intellectually satisfactory explanation of the nature of the old regime was provided by the Comte de Montlosier in his *De la monarchie française depuis son établissement jusqu'à nos jours* (4 vols., Paris, 1814). This is not an easy book to read. It is relentlessly abstract, concerned throughout with generalities and often long-winded. Moreover it is exclusively political in its view, though the author interprets the term in a wide sense, to include political psychology and to some extent ideas. Nevertheless it is a work of real analytical power and was certainly the most convincing attempt hitherto made to come to grips with the historical problems presented by the old regime in France. For Montlosier the salient characteristic of eighteenth-century France was the increasing weakness, poverty, division, and, above all, uselessness of the nobility. Excluded from civil functions, given every opportunity to squander money and none to make it, split by increasingly sharp distinctions between families of ancient lineage and those of recent ennoblement, it no longer served any useful function in society or the state. 'At this period,' he wrote, 'no one any longer knew what to do with the French nobility. It embarrassed some and offended others: it was in everybody's way.'[2] This was, however, merely one aspect of the increasing incoherence of French government in the eighteenth century. The country had no real constitution or public law. More and more it was dominated by a multitude of conflicting institutions and competing privileges between which the government had to manœuvre as best it could. France had been reduced to a state in which real change was impossible because of the opposition it would

[1] i. 143.
[2] i. 305.

arouse: any effort of this kind was likely to produce a crisis.[1] Under these circumstances the system of government had become 'weakness, varnished over with an appearance of despotism'.[2] At bottom Montlosier was a romantic conservative. His hankering after a powerful nobility, which led him to attack Turgot for his 'violent and revolutionary' attempt to abolish feudalism;[3] his belief that the Middle Ages had shown the true French character at its best, which led him to disapprove of the prohibition of duelling;[4] his repeated unfavourable contrasting of the *'parlements* of lawyers' of the eighteenth century with the *'parlements* of barons' which he believed had been their medieval predecessors: all these were exhibitions of romantic nostalgia rather than serious efforts at historical analysis.[5] But in his clear view of the widening gulf between myth and reality in many aspects of French government during the eighteenth century, and above all in his grasp of the extent to which the monarchy was limited in what it could achieve by an increasingly chaotic and complex structure of privilege, he showed more understanding than any of his contemporaries of the most intractable of the problems faced by the old regime.

Montlosier was a writer of true originality, difficult to fit neatly into any historiographical pigeon-hole. An explanation of the fall of the old regime of still greater originality, though a historiographical curiosity, and one merely outlined rather than developed in any detail, was provided by the revolutionary leader, Antoine Barnave, in his *Introduction à la révolution française.*[6] This book was justly praised a century later by Jean Jaurès, the greatest figure in the history of socialism in France, as providing a 'first sketch' of historical materialism. Barnave placed the Revolution in the widest possible historical context; all the previous history of France, even of Europe, had gone to create the situation in which it became possible. Industry, trade, the growth of towns, the diffusion of education, had all helped to produce by the accession of Louis XVI a situation in which a forcible assertion of popular power against a nobility still based

[1] ii. 34–6, 98 ff., 150–3.

[2] ii. 162.

[3] ii. 176 ff.

[4] ii. 79–96.

[5] For his belief that the future well-being of France depended upon the restoration of the power of corporate bodies, above all the nobility, see iii. 391–2, 403.

[6] First published in a rather unsatisfactory edition by Berenger de la Drôme in 1843; I have used the modern one edited by F. Rude (Paris, 1960: Cahiers des Annales, 15).

essentially on the ownership of land was likely. Social conflict might still have been prevented had the king and his ministers adopted the correct policies (for Barnave was not a complete determinist; material circumstances to him provided a field of possibilities for thought and action but did not prescribe rigidly the march of events). However, the privileges of noble birth were reasserted, the *parlements* recalled, the Third Estate increasingly excluded from military careers; thus the laws were placed in opposition to 'the natural progression of things', with inevitably fatal results. The book, unpublished for half a century after its composition and in any case no more than a hasty sketch, had no influence on historical writing. But it deserves, indeed demands, a brief mention as the boldest and most original of all contemporary analyses of the nature of the old regime and the origins of the Revolution.

The quality of the historical writing which has so far been discussed was not, by and large, very high. The atmosphere of the revolutionary and Napoleonic period was not favourable to good work of this kind. Quite apart from the political tensions and convulsions of these years, the bitter party conflicts, and the lack of distance from the subject, there was the difficulty that the Revolution found its inspiration in the present and the future rather than in what had been. In so far as it had a historical model it was republican Rome, certainly not the French past, saturated in monarchist feeling and tradition.[1] Under Napoleon the position changed considerably; but the fascination with Roman models still continued and the intellectual atmosphere of the Empire was hardly favourable to disinterested historical inquiry. In different ways the situation under the Restoration and to a lesser extent under Louis Philippe was almost equally unfavourable to the production of good history. Intellectual life became freer than under Napoleon; but all discussion of the old regime was still inevitably carried on under the over-powering shadow of the Revolution. Eighteenth-century France tended to attract attention, as before, as the seed-bed from which the Revolution had sprung rather than in its own right. Moreover political passions were still a very powerful distorting influence; and the tools essential for the drawing of a complete and fair picture of the old regime—printed documents or adequate access to un-printed ones, specialized regional and local studies—simply did not

[1] P. Stadler, *Geschichtsschreibung und historisches Denken in Frankreich, 1789– 1871* (Zürich, 1958) pp. 42, 45.

as yet exist. The publishing of historical materials which was a marked characteristic of the intellectual life of France (and of other European states) from the 1820s onwards for long did little to ease study of the old regime. The *Histoire parlementaire* of Buchez and Roux (1834–8), the reprinting of the *Moniteur* from 1840 onwards, provided at least some of the essentials for systematic discussion of the political and Parisian aspects of the Revolution. The *Collection de documents inédits sur l'histoire de France*, which began to appear in 1835 (and which included a series on the Revolution) did much to place the history of medieval France on more solid foundations. But the eighteenth century received very little attention of this kind. The result was that discussions of it continued for long to be either attempts at literature or, more commonly, political polemics or some mixture of the two.

Liberal writers after 1815 defended the Revolution, or at least its earlier and more moderate period in 1789–91, by arguing that it was merely the culmination of struggles for political liberty and representative institutions which could be traced throughout the whole history of the old monarchy. The book of Mme de Staël which has already been mentioned[1] is the best-known example of this line of thought. The argument was sometimes even widened to present the Revolution as the climax of a European, not merely French, movement for liberty, which had taken a religious form in the sixteenth century with the Reformation, a constitutional one in the seventeenth with the English civil war, and an intellectual one in the eighteenth.[2] Conservatives on the other hand minimized the social and political conflicts which the liberals thought they had discovered in the old regime and continued to deny that its collapse had been inevitable. To them the Revolution remained the result either of the malevolence of a small number of men who had plotted the events of the later 1780s or of the corrupting and disintegrating effects of the Enlightenment.[3]

But none of this amounted to even an attempt at serious analysis of the realities of eighteenth-century France. The largest work on the subject produced under the Restoration, the *Histoire de France pendant le dix-huitième siècle* of Jean-Charles-Dominique de

---

[1] See above, pp. 8–9.

[2] On liberal historical writing under the Restoration see S. Mellon, *The Political Uses of History: A Study of Historians in the French Restoration* (Stanford, 1958), chap. ii, *passim*.

[3] Ibid., chap. iv, *passim*.

Lacretelle (Paris, 1819) was the product of a professional writer and had, in a rhetorical *Académie française* style, some literary merit. Moreover as the work of a man who had been a royalist during the Revolution and its severe critic after Thermidor it is strikingly moderate and balanced in tone. Though he criticized the superficiality of the Enlightenment and the unreality of many of its hopes and expectations Lacretelle's verdict on it was by no means totally unfavourable. He understood that the eighteenth century had seen great intellectual progress in many fields and that even those *philosophes* who had the widest intellectual range—Turgot for example— were not necessarily dilettanti.[1] His estimate of the abilities of many of the ministers of the period—Choiseul, Terray, Calonne—was just and balanced. But the book is unmistakably the work of a littérateur rather than of a historian; it describes and narrates, but it does not analyse. The France with which it is concerned is that of Paris and the court; and even within these limits it is the striking incident, the intrigue, the scandal, rather than the truly significant event which attracts the author's attention.[2] The life of the provinces, social and economic developments, even the deeper aspects of intellectual life, receive virtually no attention whatever in this long book.

The same almost exclusively political and narrative character is only too visible in the relevant volumes of the *Histoire des Français* of J. C. L. Sismonde de Sismondi.[3] Sismondi himself admitted that he was much less interested in the modern than in the medieval parts of his enormous work: this lack of interest and the narrowness of his view of history emerge clearly in the meagre range of materials on which these large volumes are based—memoirs; some printed correspondence, above all that of Voltaire; a few narrative accounts, among which the mediocre productions of Soulavie are notably prominent. Even within the field of politics it is only the court and the ministers which attract attention. The provinces and the great administrative machine which really ran the country escape notice altogether, whereas wars and foreign policy are dealt with in often tedious narrative detail. The brief treatment of eighteenth-century

[1] iv. 122, vi. 283–4, iii. 91–2.

[2] For example, the long account of the attempt by Damiens in 1757 to assassinate Louis XV and his punishment (iii. 268–84), and that of the Diamond Necklace scandal in 1785–6 (vi. 115–30).

[3] Vols. xx and xxi (Brussels, 1844). Vol. xxi was in fact not the work of Sismondi but of Amedée Renée.

France in Guizot's *Histoire de la civilisation en Europe*[1] is even more exclusively focused on politics but in a different way. Whereas both Lacretelle and Sismondi are narrative and anecdotal in their treatment, Guizot is interested only in the widest generalities, the most sweeping movements of thought and feeling. For him the eighteenth century was dominated in France by a struggle between the increasingly dominant spirit of free inquiry, which was purely speculative and divorced from practical life, and the opposing force of absolute monarchical power, perfected by Louis XIV. The outcome of this struggle, the Revolution, he regarded as in some sense the heir of the English Revolution of the seventeenth century, which had been inspired by the same conflict.[2] The ultimate basis of any secure political structure must be balance between opposing forces and tendencies; eighteenth-century France had suffered first from the destruction by Louis XIV of all independent institutions capable of opposing the monarchy and then from the increasingly uncontrolled power of the Enlightenment, a power which by the end of the century had made it intolerant and led it into 'error and tyranny'.[3] Like Mme de Staël, another liberal with a marked authoritarian streak, Guizot used the history of France during the eighteenth century not as an object of serious study but as a pretext for the expression of his own dogmatisms.

The most satisfactory study of the later years of the old regime hitherto published was probably the first volume of F. X. J. Droz, *Histoire du règne de Louis XVI pendant les années ou l'on pouvait prévenir ou diriger la révolution française* (3 vols., Paris, 1839–42). Though the author was not a trained historian (his other publications are works of popular philosophy) he made a considerable effort to establish the facts as accurately as possible. As a political narrative of the events of the later eighteenth century the book was superior to its competitors. But it was still no more than a political narrative, centred like the others on Paris, the court, the monarchy and its ministers. No substantial account of the old regime had as yet been written which took more than a mainly political and narrowly Paris-centred view of its subject. In none as yet did the French people, its sufferings, feelings, and ambitions, figure as a force of significance.

---

[1] I have used the English translation, *General History of Civilization in Europe, from the Fall of the Roman Empire till the French Revolution* (Edinburgh, 1848).
[2] p. 240.
[3] pp. 237–8, 240–1.

This position was now about to be changed, by the populist rhetoric of Michelet, by the insight of de Tocquville, and, in the long run most fundamentally of all, by the exploitation of new sources of information and the progressive accumulation of knowledge.

ROMANTICS AND IDEOLOGUES: FROM MICHELET TO TAINE

In the works of Jules Michelet the people enters the historiography of the old regime for the first time; indeed from being ignored and disregarded it comes at a stroke to play the leading role. As the son of a poor printer who had himself worked with his hands in early life Michelet might have been expected to be readier than his predecessors to make the common man the hero of his story. Yet it was only after the revolution of 1830 that the history of France became his primary interest. Once this had happened, however, that interest, and the nationalist and populist emotions which underlay it, became obsessive. In 1831–44 he travelled extensively throughout France; he had undoubtedly a wide and direct knowledge of the country and its people. In 1833 he began publication of his great *Histoire de France*, of which two volumes, *Louis XV* and *Louis XV et Louis XVI*, deal with the period 1715–89.[1] In 1847 appeared the first volume of his *Histoire de la révolution française*, which began with a brief introduction on the eighteenth century.[2] The ideas which inspired his writing are best summed up, however, in his *Le Peuple* (1846), which can be regarded as a kind of confession of faith.[3] This was above all a plea for greater national unity and a more effective national life. The growth of mechanical forces and mechanistic attitudes in all aspects of life must be resisted; and this could best be done by relying on the great reservoir of healthy emotions and true idealism inherent in the people at large. Class distinctions

---

[1] I have used the undated edition published by Flammarion.

[2] I have used the edition published, again without date, by A. Le Vasseur.

[3] It is usefully summarized in G. Monod, *La Vie et la pensée de Jules Michelet* (Paris, 1923), ii. 203–8. There is a large literature on Michelet. See G. Monod, 'Michelet et l'histoire de la révolution française', *Revue internationale de l'enseignement*, i (1910), 414–37; A. Chabaud, *Jules Michelet: son œuvre* (Paris, 1929); O. A. Haac, *Les Principes inspirateurs de Michelet* (Paris, 1951); J.-L. Cornuz, *Jules Michelet. Un aspect de la pensée réligieuse au XIXe siècle* (Geneva–Lille, 1955). For critical comment see P. Lasserre, *Le Romantisme français* (Paris, 1907) and in particular P. Geyl, *Debates with Historians* (Groningen-The Hague, 1955), chap. iv.

must be swallowed up in patriotism and democracy, both of which must also be carefully inculcated in the young. Cosmopolitanism, and in particular the Catholic Church as the embodiment of this dangerous feeling, must be distrusted and resisted. France was the spiritual leader of the world, the universal fatherland, the living embodiment of the idea of fraternity which was the greatest of all the benefits brought by the Revolution.

These attitudes, with their generosity of feeling and genuine width of ideas on the one hand, their scarcely-concealed xenophobia and inflated rhetoric on the other, underlie all Michelet's major historical writings. Throughout the two volumes on the eighteenth century the same complaint continually recurs; French policy was controlled by foreign states, Britain or Austria, who unscrupulously distorted it and exploited France for their own purposes. In particular Louis XV systematically sacrificed the interests of France to those of the dynasty and especially to his desire to obtain a throne for his son-in-law, the Infante Don Philip. This meant a continual *conspiration de famille* against the country and a predominance of Spanish, and above all Austrian, interests over French ones. Thus Fleury deliberately sabotaged the French war effort in the autumn of 1741 by failing to ensure a French advance on Vienna, while the Diplomatic Revolution of 1755–6 was largely the product of the desire of Louis XV to obtain the Austrian Netherlands as a kingdom for his son-in-law.[1] In 1761 Choiseul (to whom Michelet is bitterly hostile) was unwilling to make peace because he was under the thumb of the Austrian court, while in 1779 the failure of the Franco-Spanish attempt to invade England was brought about by the unwillingness of Louis XVI and his ministers to press matters to a conclusion against France's greatest enemy.[2] Throughout both volumes the same note is repeatedly struck. The old regime was corrupt, weak, and, above all, unpatriotic. The Revolution was therefore (though this is implied rather than explicitly stated) necessary and inevitable. Repeatedly Frederick II, efficient, austere in his personal life, the Enlightenment enthroned, is used as a foil to exhibit in darker hues the defects of the French monarchy.[3]

---

[1] *Louis XV*, pp. 177–8, 288 ff.   [2] *Louis XV et Louis XVI*, pp. 81, 216.
[3] *Louis XV*, pp. 12–13, 156, 185 ff., 295–7, 335 ff., 347 ff., 355–6. This ostentatious admiration of Frederick II was a weapon later used against the old monarchy by other republican writers, e.g. A. Jobez, *La France sous Louis XV* (Paris, 1864–73), v. 12, vi. 290–4.

These volumes do not show Michelet at his best; the eighteenth century could never inspire him as the Revolution did. They have the rhetorical bite, based on sweeping assertions expressed in short sentences, of everything he wrote. They bear the imprint of a distinctive personality as the work of Lacretelle or Droz does not. But they also have the defects of marked individuality. They are poorly organized and repeatedly wander into byways which interested Michelet personally; this makes for a marked lack of balance. For example they contain a fairly long discussion of the ideas of the moralist Vauvenargues,[1] while those of Montesquieu, arguably the greatest political and social thinker of all modern history, are by contrast totally neglected. There is a marked anecdotal tendency, a lingering over picturesque incidents sometimes out of all proportion to their intrinsic importance.[2] In spite of Michelet's claims to have used archive materials (as he undoubtedly did for his work on the Revolution) the footnote references are virtually entirely to printed memoirs, well-worn sources of often mediocre value. Above all the volumes suffer, in spite of the author's populism, from the same limited field of vision, the same concentration on Paris, the court, the ministers, the same indifference to the provinces and to economic and social forces, which had afflicted all his predecessors. Nevertheless with Michelet a new and seductive spirit enters discussion of the old regime; the case for its political sterility and the necessity of its downfall had now been made with unprecedented rhetorical skill and popular appeal. In France the romantic populism and nationalism which inspired him found an echo in the huge *Histoire de France* of Henri Martin.[3] Across the Channel a spirit in some ways similar had already inspired the discussion of the old regime in the first pages of Thomas Carlyle's *The French Revolution: A History* (London, 1837).

Martin represented the official values of the early years of the Third Republic rather as Guizot had done those of the July Monarchy; his status in this respect was recognized when on his

[1] *Louis XV*, pp. 361–6.

[2] For example the discussion of the alleged stigmatization and later pregnancy (allegedly by a Jesuit) of Mlle Cadière, which caused a popular sensation in 1731 (*Louis XV*, pp. 101–10); the account of the Damiens incident (*Louis XV*, chap. xix); the very long account of the Diamond Necklace scandal (*Louis XV et Louis XVI*, chaps. xvi–xviii).

[3] The first edition appeared very rapidly in 1833–6. I have used the revised and improved 4th edition (Paris, 1855–9).

death in 1883 he was given an official funeral.[1] Like Michelet he was above all a nationalist historian; he too attacked the old regime for its readiness to sacrifice French interests to those of the ruling family and of foreign powers, England and Austria.[2] Like Michelet he was a romantic historian; but much more blatantly than his great predecessor he exemplifies the dislike of superficial rationalism, the hankering for faith, for well-grounded traditions and a clear line of descent uniting the present with the past, which marked much romantic historiography. The most interesting general form taken by these feelings, in Martin's case, was his belief in the importance of the Celtic contribution to French history and the French mind. In his treatment of the eighteenth century it is seen in his dislike of Voltaire and preference for Montesquieu, and in his attacks on the *philosophes* for their destruction of so many of the beliefs of the ordinary man— in free will, the immortality of the soul, the importance of the conventional virtues—and their failure to provide him with anything which adequately replaced these lost certainties.[3] Like Michelet again he was a passionately republican writer, unsparing in his denunciations of the corruption and incompetence of the old monarchy in its last decades. For a few years after 1715, he thought, the Regency had at least been marked by originality and by a great social experiment, the Law system, which had been in many ways beneficial to the country as a whole. The reign of Louis XVI had seen the Enlightenment attempt unsuccessfully (he is thinking here above all of Turgot) to save the country from the strife which was to come. But the reign of Louis XV had been a period of complete governmental torpor; and like a good bourgeois republican he was deeply shocked by the intrigues and immorality of the court during much of the eighteenth century, by a display of vice which seemed to him to recall the Byzantine Empire or some Asiatic despotism.[4] If Martin was in some sense in the tradition of Michelet, however, he completely lacked Michelet's very real gifts as a writer; and his view of his subject was if anything less wide than that of his great predecessor. Nearly a quarter of the two large volumes he devoted

---

[1] For a general discussion of his intellectual and political significance see C. Rearick, 'Henri Martin: from Druidic Traditions to Republican Politics', *Journal of Contemporary History*, vii (1972), 53–64. The best biography is G. Hanotaux, *Henri Martin* (Paris, 1885).

[2] *Histoire de France*, xv. 122.

[3] xv. 392, 408–9, xvi. 58–9.

[4] xv. 2, 71–3, xvi. 669–70.

to the France of 1715–89 was concerned with intellectual life; but the rest is merely the narrative of politics, the court, diplomacy, and war which was still, when he wrote, the only kind of history that most readers were willing to take seriously.

Carlyle is also, if the shop-soiled adjective has any precise meaning, a romantic historian. The similarities between him and Michelet are often striking and it is possible, though quite unproved, that Michelet was inspired by him.[1] They resemble each other in their styles, pungent, highly individual, and making a powerful appeal to the emotions. Above all they resemble each other in that they see the people as the real hero of their story and the selfish disregard of the people's sufferings and grievances as the weakness which eventually destroyed the old regime. Both saw history as a kind of poetry; and to both the poem was in the last analysis written by God.[2] To both history was above all a tool with which to expose the essentially moral nature of the world. 'Wherever huge physical evil is,' claimed Carlyle, 'there, as the parent and origin of it, has moral evil to a proportionate extent been.'[3]

There are, however, significant differences between the two writers. In particular Carlyle is much readier than Michelet to endow individuals with a symbolic or allegorical significance which has no necessary connection with their real personalities or abilities; thus to him Louis XVI symbolizes the weakness and failure of the old regime even when it meant well. Moreover he is much more hostile than Michelet to the Enlightenment. He expresses with great force the by now well-established conservative criticism of it as superficial, purely analytical, and therefore unconstructive, tending to loosen the bonds which held society together and to weaken the beliefs which had given meaning to men's lives. He complained of the

---

[1] Carlyle's book appeared ten years before the first volume of Michelet's *Histoire de la révolution française*; but there is no evidence that Michelet read Carlyle in English and the French translation was published only as late as 1865–7 (A. Aulard, 'Carlyle historien de la révolution française', *Études et leçons sur la révolution française*, 7ème série (Paris, 1913), p. 198). On *The French Revolution* in general see Louise M. Young, *Thomas Carlyle and the Art of History* (Philadelphia, 1939) and C. F. Harrold, 'Carlyle's General Method in "The French Revolution"' *Proceedings of the Modern Language Association*, xliii (1928), 1150–69.

[2] See on Carlyle in this respect the comments of H. M. Leicester Jr., 'The Dialectic of Romantic Historiography; Prospect and Retrospect in "The French Revolution"', *Victorian Studies*, xv. No. 1 (Sept. 1971), 5–17.

[3] *French Revolution* (1898 edn.), i. 36.

eighteenth century that 'Faith is gone out; Scepticism is come in', and spoke contemptuously of 'this perpetual theorising about Man, the Mind of Man, Philosophy of Government, Progress of the Species and such-like'.[1] All this added up to a depressing picture of eighteenth-century France. To Carlyle it seemed without a redeeming feature, a desert of the human spirit made productive only by the fertilizing cataclysm with which it ended. No writer has seen the old regime more uncompromisingly as a mere setting for and prologue to the Revolution. The solid economic progress achieved, the vast constructive possibilities of the Enlightenment, even the great virtues of the administrative machine which held the country together and of many of the men who ran it; none of these meant anything to him. There is no minister of the last decades of the monarchy for whom he has sympathy: not for Terray ('dissolute Financier, paying eightpence in the shilling'); not for Maupeou, with his 'sinister brows' and 'malign rat-eyes'; not for Calonne, a mere talker and hence worthy only of the deepest contempt; not for Loménie de Brienne ('let us pity the hapless Loménie; and forgive him; and, as soon as possible, forget him').[2] For him the fall of the old regime had nothing to do with impersonal forces of any kind; it was the result of wrongdoing by individuals. The nobility indulged in 'debauchery and depravity . . . perhaps unexampled since the era of Tiberius and Commodus', while Louis XV on his death-bed is subjected to a stream of moralizing reproaches—'thy foul harem; the curses of mothers, the tears and infamy of daughters', and so on.[3] Carlyle paints a picture whose vividness has never been surpassed; and even on the technical level, in terms of the width of his reading and the critical ability he displays in his use of sources, he has won the praise of professional historians well qualified to judge.[4] But his arbitrary judgements and arrogant moralizing closed the door to new ideas and were at bottom profoundly backward-looking. There is no great figure in historiography whose work is more clearly a blind alley.

Michelet and Carlyle are two of the summits of the historical writing of the mid-nineteenth century. The third, of quite a different

[1] i. 14, 53. cf. 54, 57–8.
[2] i. 3, 4, 66–9, 110.
[3] i. 12, 20.
[4] Aulard, 'Carlyle historien', pp. 199–203. One of the best of the many general discussions of him as a historian is P. Geyl, *Debates with Historians* (Groningen–The Hague, 1955), chap. iii.

kind and intellectually towering above both, is Alexis de Tocqueville.[1] His *L'Ancien Régime et la révolution* (Paris, 1856) is striking above all because of the freshness of its approach; no more original work of history was published during the nineteenth century.[2] On the methodological level it shows its originality in being the first important work on the old regime to make serious use of the enormously rich archival materials accumulated during the eighteenth century in the French provinces. De Tocqueville spent a year in Tours and studied intensively the archives of the Indre-et-Loire departement, realizing that they provided insights and information which were not to be had from the materials available in Paris. On a higher level the book differs from almost all its predecessors in being an analysis, not a story or a narrative. It has a questioning and reflective aspect which makes de Tocqueville resemble Burke or Montesquieu (both thinkers with whom he has, in different ways, something in common) rather than any historian of the first half of the nineteenth century. Yet with all its great virtues the book was the product of a profound pessimism. De Tocqueville, a member of an old, though not particularly wealthy, Norman noble family, had grown up in an atmosphere dominated by memories of 1789 and the following years. He had himself seen the revolution of 1830, the much greater one of 1848, and the *coup d'état* of 1851 which paved the way to the creation of the Second Empire. He was thus dominated in his last years by a passionate consciousness that France had totally failed to achieve either the deeply-engrained feeling for liberty or the political stability both now so characteristic of England (which, like so many French liberals of his age, he regarded as a constitutional model). Why was this so? His book was an effort to answer the question. The old monarchy, he argued, had undermined and by 1789 largely destroyed the real power of the nobility,

[1] There is a large and still growing literature on de Tocqueville. The best general study of his work on the old regime is R. Herr, *Tocqueville and the Old Regime* (Princeton, 1962). See also M. Reinhard, 'Tocqueville, historien de la Révolution', *Annales historiques de la révolution française*, xxxii (1960), 257–65; G. Lefèbvre, 'A-propos de Tocqueville', ibid., xxvii (1955), 313–23; and H. Brogan, 'Alexis de Tocqueville; the Making of a Historian', *Journal of Contemporary History*, vii. Nos. 3–4 (July–Sept. 1972), 5–20. The most recent publication of this kind is E.-F. Guyon, 'Tocqueville et son destin', *Revue d'histoire diplomatique*, Oct.–Dec. 1972, 289–307.

[2] I have used, as the edition likely to be most accessible to the English-speaking reader, the paperback Fontana reprint (London, 1966) which has an introduction by Hugh Brogan.

and also autonomy and self-determination in town and village government, by a sustained policy of centralization. It had even interfered with the ordinary functioning of the judicial system, notably by the technique of 'evocation' of cases to the king's council.[1] The Council, the intendants, the ministers such as the Controller-General, had all contributed to these processes.[2] The loss by the nobility of their former civil and military functions, their reduction to uselessness, had made the privileges which they retained seem to the mass of the population increasingly intolerable. Indeed these privileges reached their apogee as the decline in the real importance of the nobility became complete. Moreover the monarchy and its ministers, in order to dominate French society more absolutely, had deliberately fostered the growth of divisions and class antagonisms within it.

The Revolution, far from marking a break with the old regime in this work of centralization, had completed it; the real objective of the revolutionaries had been to destroy yet more completely any institution or tradition which stood in the way of their untrammelled power.[3] By completing the destruction of the old aristocracy and implanting in France democratic political structures they had introduced a new and terribly dangerous element of instability and arbitrariness into French life. For the existence of an effective aristocracy, based on the ownership of land and independent of the central government, was the essential conditon of civil and political liberty. This belief was fundamental to de Tocqueville's thought; his book is a sustained exercise in nostalgia for the days when the class to which he belonged by birth had been able to restrain the central government and thus preserve some element of autonomy and self-determination in French life. Though he fully accepted the inevitability of some basic equalities between the different social classes in a modern state—for example before the law and as regards access to official positions—he was unquestionably an élitist, the greatest of modern times. His contempt for fortunes gained in trade and industry, his belief in landownership as the only real guarantee of political independence, above all his deep distrust of the masses, of their passions, their intolerance, their willingness to sacrifice liberty to the gratification of material desires, all illustrate this.

It is not difficult for the present-day historian to find gaps and

---

[1] pp. 70–6, 76 ff., 80 ff.   [2] pp. 61 ff., 64.
[3] pp. 60–1, 111, 131, 158, 40, 207–11.

flaws in the picture painted by de Tocqueville. He was not interested in economic questions. He assumed too readily that the difficulties and hardships of the peasantry were the results merely of the survival of feudal tenures and the manorial system; he knew nothing of the 'feudal reaction', in effect a commercialization and distortion of feudalism, which most modern historians have detected during the last decades of the old regime. For him the agrarian picture was dominated by the small peasant landowner; other groups such as the tenant farmer and the agricultural labourer, whose difficulties were more purely and brutally economic, received little of his attention.[1] Above all he was unfair to the French monarchy. His view of it as hungering for power for its own sake ignored completely the fact that France had to maintain her position in a competitive state-system and that for this a strong monarchy, acting through a large bureaucracy and powerful armed forces, was essential.[2] As Professor Herr has pointed out, it was only after the war of 1870 had brought home to France as never before the penalties of military and diplomatic weakness that a realistic estimate of the place of foreign policy in the structure of the old monarchy was made. Such an estimate was achieved above all by Albert Sorel in the first volume, published in 1885, of his eight-volume *L'Europe et la révolution française*. Just as de Tocqueville showed that in some important ways at least the revolution had continued the work of the old monarchy in the government of France, so Sorel showed that in foreign policy the revolutionaries had acted in most respects in the monarchical tradition, pursuing objectives for which Richelieu and Louis XIV had striven.[3]

---

[1] Though he was aware of the desirability of a detailed investigation of land-ownership in the French countryside under the old regime (pp. 54–5).

[2] For a good illustration of the unfairness of his attitude in this respect see p. 239.

[3] There is a contemporary foreshadowing of Sorel's argument in the Comte d'Hauterive's *De l'état de la France à la fin de l'an VIII* (Paris, 1800). More or less simultaneously with Sorel's work, André Chuquet showed that the great military achievements of the Revolution had their roots in the reforms and improvements carried out in French military organization during the last decades of the monarchy (A. Chuquet, *Les Guerres de la révolution* (11 vols., Paris, 1886–96)). A completely revolutionary army was thus shown to be as much a myth as a completely revolutionary foreign policy or administrative structure. In the same year that Sorel's first volume appeared René Stourm in his *Les Finances de l'ancien régime et de la révolution* (2 vols., Paris, 1885) showed that many of the financial reforms of the Revolution had their roots in the improvements of the reign of Louis XVI, when most of the foundations for the French financial system of the nineteenth century had been laid (i. 477–86, 501).

But no criticism of this kind can reduce more than marginally the value of de Tocqueville's achievement. He broke decisively with the convention which discussed the old regime and the outbreak of the Revolution largely or mainly in terms of personalities. (Michelet had also done this, but in a visionary and incantatory, not an analytical way; he had none of de Tocqueville's desire to make history resemble, in its tone and methods, one of the physical sciences.) With *L'Ancien Régime et la révolution* a great historian discussed a great historical question for almost the first time in terms of social classes. 'I speak of classes', wrote de Tocqueville, 'they alone should concern history.' This remark in itself marks him as a truly modern historian, one with whose methods and assumptions a present-day practitioner of the art can feel at home as he cannot when reading Lacretelle, Guizot, Michelet, or Carlyle. More important still, he struck a great blow at the simplistic belief that the Revolution had changed everything, that it meant, for good or evil, a complete break with the past. After de Tocqueville it was much easier to see it as most present-day historians do—as one step, though a long one, in a series which leads from the France of the Middle Ages to the France of today. The naïve Enlightenment belief in the possibility of a people disowning its past and making a completely new start, a belief which had survived into the nineteenth century, was slowly but increasingly to give way to a more realistic view of societies as inevitably to a large extent the prisoners of their pasts; in this transformation de Tocqueville played a leading role.

His book was a powerful and brilliant expression of a small number of basic ideas; for all its scholarship it was a profoundly ideological work. The same is true of a number of less important discussions of the old regime published in the middle decades of the nineteenth century, nearly all of whose authors, like de Tocqueville, wrote still under the influence of the Revolution and in response to the challenge it offered. The most interesting and original of these minor ideologues is the socialist Louis Blanc; the first two volumes of his *Histoire de la révolution française*, which deal with the *ancien régime*, were published in 1847. Blanc detected throughout history the working of three great conflicting principles—authority, which meant coercive power, inequality, and a superstitious respect for tradition; individualism, which meant *laissez-faire* in government and, in life generally, the assumption that the individual was sole judge of himself and his environment; and fraternity, which meant regarding societies as organisms on the age-old analogy of the human body and basing government on persuasion and the assent of the governed. The eighteenth century in France, he argued, had seen a growing current of individualism, represented in differing ways by Voltaire, Montesquieu, and Turgot, and a growing freedom and power of the bourgeoisie; this had culminated in the middle-class Revolution of 1789 typified by Mirabeau. Side by side with this, however, had developed a counterbalancing movement towards fraternity, whose prophets were Rousseau, Mably, and, surprisingly, Necker, and which underlay the radical phase of the Revolution from 1792 onwards typified by Robespierre.[1] *Laissez-faire*, selfish and anarchical, must be resisted; man could progress only in association with others, through fraternity.

This faith emerges clearly in his book, sometimes in an engagingly dogmatic form. Thus he was sympathetic to the ideas and policies of Law, partly because he correctly saw that these had helped to shake old ideas and habits but also because he believed that paper money, whose basis was confidence, was the natural currency of a society based on fraternity and 'association'. Gold and silver coins by contrast were the currency of an individualist society based on competition and distrust.[2] Repeatedly he attacks the Enlightenment for its excessive intellectualism, its contempt for feeling, its tendency to see the moral dignity of the individual only in his isolation from his fellows. Neither Montesquieu nor the physiocrats had anything to offer the ordinary man, who needed not a sterile freedom but effective protection in an increasingly competitive society.[3] Rousseau on the other hand he praised lavishly since 'amid the apostles of individualism he thought like the Nazarene', and because he opposed to the cult of reason which divided men from one another the cult of sentiment which united them.[4] Though he had some admiration for Turgot he deeply disliked his ideas as tending to put the poor at the mercy of the rich and to exalt individualism through the establishment of economic freedom. Terray, on the other hand, he admired for his ruthless efforts in the early 1770s, as Controller-General, to improve

[1] This summary is based on E. Renard, *Louis Blanc: Sa vie—son œuvre* (Paris, 1922), pp. 194–5; for another general account see L. Loubère, 'The evolution of Louis Blanc's political philosophy', *Journal of Modern History*, xxvii (1955), 39–60.

[2] *Histoire de la révolution française*, i. 271–306, especially p. 278.

[3] i. 458, 525–6.   [4] i. 397–403.

the financial position of the monarchy, and for his willingness to override the interests of the rich and powerful to this end (an attitude rare in the increasingly property-conscious nineteenth century).[1]

Blanc's ideas had no real future. Well before the end of his long life (he died in 1882) it had become fairly clear that Marxism was the wave of the future in European socialism; but he was no Marxist and was quite explicitly anti-materialist. 'It is not force which controls the world', he wrote. 'Whatever appearances may say, it is thought; and history is made by books.'[2] Nevertheless in some ways his work seems remarkably modern. It has obvious weaknesses. There is surprisingly little, considering his basic assumptions, on economic history or society; the emphasis is overwhelmingly on Paris, the court, and the central government, while the provinces are grossly neglected. He used only printed materials, mainly memoirs, and was uncritical in his use of them. Yet he is fair in his judgements and is always striving to understand as distinct from merely narrating. In one way in particular he points to the future. He was the first writer to grasp at all fully the strength of the anti-intellectual and anti-rational tendencies in French intellectual life during the decade or more before the Revolution. The reaction against the Enlightenment to be seen in mesmerism, Freemasonry, illuminism, and martinism, the hankering for the comfort of belief as well as for the pride of knowledge which it expressed, had not altogether escaped earlier writers.[3] But Blanc more than any of his predecessors grasped its importance in the intellectual climate of the 1780s, even though his account of it was often uncritical.[4]

Neither the disillusioned élitist liberalism of de Tocqueville nor the humane utopian socialism of Blanc were typical of the ideological approach to the old regime which was most dominant in French writing in the middle decades of the nineteenth century. This continued to be a republicanism now hardening into what was to become, under the Third Republic, an official orthodoxy.[5] Henri

---

[1] i. 526–31, 467.   [2] i. 447.

[3] See e.g. Lacretelle, vi. 97–103 for a fairly detailed account of the mesmerist episode.

[4] He discusses it in vol. ii, chap. iii. He believed, as an illustration of his lack of critical sense, in the existence of a great network of spies and informants which had in the 1780s given the illuminist secret societies news of events everywhere in Europe.

[5] On the development of this orthodoxy see the hostile but in some ways penetrating account in D. Halevy, *Histoire d'une histoire, esquissée pour le troisième cinquantenaire de la révolution française* (Paris, 1939).

Martin is the best known representative of this attitude. Another, more intelligent and more extreme, is Pierre Lanfrey, in his *L'Église et les philosophes au dix-huitième siècle* (Paris, 1855) and *Essai sur la révolution française* (Paris, 1858). To Lanfrey the eighteenth century in France was above all the period which saw an unprecedented effort to assert the intellectual freedom of the individual and the free play of the mind against the intolerance and traditionalism of the Catholic Church. From the 1750s onwards there had taken place 'the most magnificent explosion of intelligence which has perhaps ever been seen'; and this had provided the indispensable basis for the political transformation at the end of the century.[1] This 'marriage of philosophy with the realities of life' had been the greatest fact of all modern history.[2] Lanfrey's books, abstract, rhetorical, and high-minded, are now of little interest in themselves. But they expressed, above all in their anticlericalism, a point of view which was already powerful among French intellectuals and which was to become more so.

More original, though in some ways not totally dissimilar from that of Lanfrey, was the positivist interpretation of eighteenth-century France provided by the English writer H. T. Buckle in the second volume of his enormously popular and successful *History of Civilization in England* (London, 1857–61).[3] Like Lanfrey, Buckle was deeply hostile to metaphysics, which to him seemed a trivial distraction from the great intellectual task of disentangling and exposing the regularities of behaviour, the predictability, which men and therefore historical events displayed. Like Lanfrey and indeed liberals generally (in this as in so many other ways de Tocqueville is an exception) he believed in the reality of progress; and this could mean only progress towards intellectual liberty. 'The real history of the human race', he asserted, 'is the history of tendencies which are perceived by the mind, and not of events which are discerned by the senses.'[4] The old regime had doomed itself by opposing and persecuting the intellectuals and writers of France, with the result that by about 1750 their criticisms had begun to be directed against political institutions and not merely against ecclesiastical ones. By the later eighteenth century French

[1] *L'Eglise* . . . (1879 edn), p. 141; *Essai* . . ., pp. 45–6.
[2] *L'Eglise*, . . ., p. 135.
[3] I have used the edition of 1878. On Buckle generally see G. R. St. Aubyn, *A Victorian Eminence: the Life and works of Henry Thomas Buckle* (London, 1958).
[4] ii. 324.

intellectual life was dominated by the idea of 'the inferiority of the internal to the external'. This meant a growing interest in the physical sciences and an increasing hostility to theories of innate ideas and to traditional religious beliefs and inhibitions.[1] These changes the old regime was completely incapable of understanding, for

. . . . if ever there existed a government inherently and radically bad, it was the government of France in the eighteenth century. If ever there existed a state of society likely by its crying and accumulated evils, to madden men to desperation, France was in that state. The people, despised and enslaved, were sunk in abject poverty, and were curbed by laws of stringent cruelty, enforced with merciless barbarism. A supreme and irresponsible control was exercised over the whole country by the clergy, the nobles, and the crown. The intellect of France was placed under the ban of a ruthless proscription, its literature prohibited and burned, its authors plundered and imprisoned.'[2]

The ideologists, the writers dominated by a belief or an assumption who have been described in the preceding pages, were all to varying extents hostile to the old regime. Whether they were dominated by the nationalism and populism of Michelet, the apocalyptic moralism of Carlyle, the pessimistic élitism of de Tocqueville, the utopian belief in fraternity of Blanc, the anticlerical intellectualism of Lanfrey or the naïve positivism of Buckle, all to some degree or another believed that the old regime had been destroyed by its own failures and defects and that the Revolution was in some sense inevitable. This idea did not go unchallenged. There were still monarchists in the middle decades of the century who stressed the virtues of the monarchy and blamed the envy and impatience of the French people and the *mauvaises leçons* of the *philosophes* for its fall.[3] More interesting and important, these offered at least one resounding scholarly defence of it in its last years, in *Les Assemblées provinciales sous Louis XVI* (Paris, 1864) by the conservative Léonce de Lavergne. All the gains of the Revolution, and more, he asserted, could have been achieved peacefully with a little patience and good sense, for 'France never enjoyed more freedom than in 1788 and 1789; instead of developing political liberty, the Revolution simply stifled it.'[4] For the protection of liberty communes and provinces must have

[1] ii. 327 ff., 361, 405.
[2] ii. 244.
[3] e.g. Comte Frédéric de Falloux, *Louis XVI* (Paris, 1840). This essentially anecdotal book had considerable success; it had gone through four editions by 1860.
[4] Preface, pp. iii–iv, x.

some genuine autonomy (here there is general agreement with de Tocqueville); and the Revolution interrupted a movement in this direction, seen in the creation of provincial assemblies in Berri, Dauphiné, and Haute-Guienne in 1778–9 and in most of the other *généralités* in 1787–8. These, Lavergne argued, constituted a noble and largely successful experiment which was never given time to show what it could achieve. 'Nothing', he concluded, 'is less like the France of Louis XIV and Louis XV, so gloomy and downcast, than this France of Louis XVI, so lively, so proud, so free, so full of hope, confidence and energy.'[1] Lavergne's work was a formidable statement, within its limits, of a moderate and well-informed conservative point of view. Nor was he the only writer of the mid-nineteenth century to show a grasp of the great constructive achievements of the old monarchy.[2] But it was above all Hippolyte Taine, in the first volume of his *Origines de la France contemporaine* (Paris, 1875–93), who provided what was to remain until our own day the most famous statement of a disillusioned conservative view of eighteenth-century France.

Taine was in many ways what is conventionally thought of as a supremely French type of intellectual. His range of knowledge and interests was remarkably wide. He was trained in philosophy and literature but became deeply interested in the sciences, above all in psychology; he was also profoundly influenced by the sociology of Auguste Comte. He thus brought to the writing of history a mind unusually well equipped, at least in formal terms. Yet he was also a Cartesian and a dogmatist. He reasoned in purely deductive terms and for all his frequently expressed desire to make history a science (the noble and pathetic ambition of so many nineteenth-century intellectuals) he had little real understanding of scientific method or willingness to allow himself to be guided by observed facts. His reading was wide, his collection of information assiduous; but he used the facts he accumulated to defend preconceived ideas which he never contemplated abandoning. These ideas were the outcome of the political events of his own lifetime. Never a supporter of the Second Empire, he was driven to a more deeply entrenched and em-

---

[1] p. 479.

[2] See e.g. the tribute paid to the great development of public works, above all roads, to the administrative improvements of the generation before the revolution, and to the efforts of the government to improve French agriculture in the second half of the eighteenth century in C. Dareste de la Chavanne, *Histoire des classes agricoles en France* (2nd edn., Paris, 1858), pp. 435–6, 439–40, 449–50, 486.

bittered conservatism by the catastrophic events of 1870–1. He despised and feared both Bonapartism and radical revolution; against them he asserted the claims and values of the well-to-do professional classes and upper bourgeoisie, highly educated, pacific, within limits genuinely liberal, but politically timid. His outlook and ideals had much in common with those of Guizot; both men represented the moderate conservative forces which had been defeated in France in the 1840s and were to be defeated once more in the 1870s.[1]

Taine's *Origines de la France contemporaine* is essentially a deeply hostile study of the Revolution of 1789 and its meaning for French history; but the first volume is an extensive analysis of the old regime.[2] It was clear to Taine that there had been much wrong with the France of the old monarchy. He was far from being an uncritical admirer of the old regime and realized very well that under it large sections of the French people had suffered real deprivations and felt real grievances. The nobility had lost its former functions and all real social utility as the scope and activity of the central government, working above all through the intendants, increased. It had thus become by 1789 a useless encumbrance of rural society, claiming dues which had once been justified but which now could only arouse increasing resentment.[3] This functionless class had become increasingly selfish. It had also become increasingly dependent on the gifts, sinecures, and pensions doled out to it by the central government whose all-pervasive activity helped to keep the provinces passive and subservient to Paris and to stifle all other public institutions. 'Bureaucracy at the centre, arbitrary power, exceptions and favours everywhere—such is the system in its essentials.'[4] A mass of evidence—their extravagance, the waste and pilfering which they allowed in their households, the loosening of family ties among them, their passion for the theatre and the opera-house (the amusements of

[1] The literature on Taine is extensive. A. Cobban, 'Hippolyte Taine, Historian of the French Revolution', *History*, lii (1968), 331–41, is a convenient summary. See also A. Aulard, *Taine, historien de la révolution française* (Paris, 1907); V. Giraud, *Essai sur Taine* (Fribourg–Paris, 1901) and *Hippolyte Taine: Études et documents* (Paris, 1928); A. Chevrillon, *Taine. Formation de sa pensée* (Paris, 1932); S. J. Khan, *Science and aesthetic judgement: a study in Taine's critical method* (London, 1953); H. Sée, *Science et philosophie de l'histoire* (Paris, 1928), pp. 383 ff.

[2] I have used the second edition (Paris, 1876).

[3] i. 36–52.

[4] i. 77–100.

those who 'played at life like children')—seemed to him to demonstrate the nobility's loss of the ability and right to lead society. The inadequacy of the traditional ruling class, the lack of local patriotism or effective local institutions, above all the stultifying and corrupting power of an overgrown central government which divided society in order to control it more effectively, meant that by 1789 there was no social unity, no common interest or feeling left in France.[1] Moreover Taine was not indifferent to the material sufferings of the poor and exploited majority of the nation. He provided extensive, though by modern standards unscientific, detail about the difficulties of the common man and particularly about the archaic and unfair taxation system under which he suffered. The appearance of the countryside and its inhabitants in France during the eighteenth century was as bad as in Ireland, he thought, while the position of the French peasant was comparable to those of India or Egypt.[2]

All this amounted to a forcible indictment of the old regime. Nevertheless these defects were superficial, or at least in principle remediable. Its collapse was the result of more deep-rooted and intractable errors. It resulted above all from a fundamentally wrong and deeply misleading view of the nature of man and human reason, and of the kind of social and political change which it was possible to carry out in human societies. Political and social structures, to Taine, are not to be decided merely by transitory human tastes and preferences. They cannot be freely chosen or subject to rapid alteration, for they are the products of the national history and character. The sudden invention of durable new institutions or a successful new constitution is beyond the powers of the human mind; such creations can be the result only of time and of a process of unconscious choice and development.[3] This is ultimately because the ordinary man, even when he is semi-educated, cannot really understand abstract reasoning or statements of general principles; to him bodily needs, his temperament (which is determined by physical factors), his instincts, traditions, imagination, and passions, are far more important. He is still in some sense an animal, brutal, violent, and greedy; and in daily life these animal appetites must be kept in check by artificial restraints, in effect by the coercive power of the state.[4]

[1] i. 166 ff., 197–203, 515–18.
[2] i. bk. v, *passim*, particularly pp. 447, 467.
[3] i. Preface, pp. ii–iii, contains a forcible statement of these ideas.
[4] i. 313–16.

The Enlightment, with disastrous results, had in its false pride and wrongheaded optimism disregarded these fundamental truths. The increasing importance of the physical sciences had helped to produce in the eighteenth century the modern idea of progress. More important, the dominance of the classical tradition and spirit had produced a marked preference in every aspect of intellectual life for the clear, the simple, the regular, the abstract. This had generated a reason which was not true reason but merely a 'raison raisonnante' which avoided contact with 'the plenitude and complexity of reality'.[1] Such an attitude involved a profoundly dangerous contempt for specialized knowledge, for intimate understanding of the irregular and intractable facts of real life and daily experience. The victory of such a spirit meant that whereas hitherto the basis of all rights had been traditional and historical henceforth they were thought of as rooted only in reason. But this reason was an incomplete and limited one which did not understand or sympathize with what it was sweeping away and did not realize that 'hereditary prejudice is a kind of unconscious reason'.[2] Montesquieu, alone among the writers of the eighteenth century in France, had had a sense of realities; but though respected he was not influential. Helvétius and d'Holbach had constructed a merely mechanical and utilitarian view of human nature and morality; Rousseau, in spite of his generosity and feeling for justice, had 'seen dreams instead of realities, lived in a romance and died under the nightmare which he had created for himself'.[3] But the agreeable way in which they were presented by many writers, and the distinctively French taste for wit and conversation, meant that the new attitudes and ideas enjoyed in France a disastrously complete diffusion which they did not achieve anywhere else in Europe.[4]

Taine's view of the old regime provides plenty of targets for criticism. His deep pessimism, his contempt for the ordinary man, his

[1] i. 250.
[2] i. 270 ff.
[3] i. 278, 282-8, 289.
[4] This type of embittered and deeply pessimistic denunciation of the Enlightenment was not peculiar to Taine. A. A. Cournot, for example, in his *Considérations sur la marche des idées et des événements dans les temps modernes* (2 vols., Paris, 1872), attacked almost simultaneously the enlightened scheme of moral values as 'an empty or false idea put at the disposal of sophists to dupe honest men' (ii. 59-60), stressed 'the vanity and emptiness of all attempts at political rationalism' (ii. 74), and argued that lasting political structures could only be organic and the work of time (ii. 82-3).

naïve assumptions about the determinants of human action and the possibility of writing 'scientific' history, have never lacked antagonists. On a more narrowly historiographical level it is not difficult to point out weaknesses in his work. There is no sense of change through time in his account of the *ancien régime*; facts taken indiscriminately from any part of the century before 1789 are lumped together in a very cavalier way to present a picture of an ideal or typical state of affairs, fixed and unchanging.[1] His picture of the monarchy as overwhelming all other political institutions is difficult to reconcile with the often bitter struggles (which he does not mention) between crown and *parlements* in France during the second half of the eighteenth century. He is notably impressionistic and imprecise in much of what he says about economic and social life.[2] His view of the Enlightenment and its leading figures is obviously distorted by prejudice and by a Cartesian determination to make the facts fit a preconceived thesis. Rousseau, for example, whom he attacks for his abstractness and lack of interest in the specific and concrete, showed a considerable grasp of the importance of local problems and particular situations in his proposals for reforming the government of Poland and for a constitution for the island of Corsica. More generally, Taine puts forward, by his insistence on the authoritarian side of the Enlightenment to the total exclusion of its more libertarian aspects, a very one-sided picture of a complex body of ideas produced by different writers over a considerable period of time.[3] Moreover he never comes to grips with the fact that the *esprit classique* which he so disliked and despised was, after all, a European phenomenon, and yet did not produce elsewhere the devastating results which he attributed to it in France. Even on the purely technical level his work can be criticized for his sometimes cavalier use of his materials and above all for the inadequacy of his documentation (though there is an inevitable element of unfairness in subjecting to modern standards of this kind any book written so long ago, especially one whose subject is so enormous).[4]

---

[1] See the criticisms in Aulard, *Taine*, pp. 31–2.

[2] e.g. the account of feudal rights at the end of the old regime in i. 25–33.

[3] This one-sided view, which relies heavily on a stressing of Rousseau's insistence on state control of education and the creation of a new civic religion, is well illustrated in i. 319–28.

[4] See the detailed technical criticisms in Aulard, *Taine*, pp. 33–9; but cf. the generally effective defence of Taine's methods in A. Cochin, 'La Crise de l'histoire révolutionnaire: Taine et M. Aulard', in Cochin's *Les Sociétés de pensée et la démocratie: études d'histoire révolutionnaire* (Paris, 1921), pp. 63 ff.

Yet when every allowance has been made for these defects Taine's picture of the old regime still remains one of the most powerful of those produced during the nineteenth century. Whatever its short-comings it provided a salutary corrective to the picture of the Enlightenment and its effects in France put forward by republican idealists such as Lanfrey; and it bequeathed to future writing on the subject a legacy which was to prove permanent. During the half-century from the later 1870s onwards, the heyday of the Third Republic, Taine's pessimism and contempt for revolutionary idealism made him an enemy in the eyes of orthodox republican historians. From the 1930s onwards historical writing about eighteenth-century France has concerned itself increasingly with issues of whose very existence he and his contemporaries were scarcely aware; the sort of book which he wrote is now almost a museum piece. Yet the attitude to the Enlightenment which he so forcibly expressed has never ceased to attract adherents. Four decades after the appearance of his volume on the old regime his intellectual heir, the brilliant conservative Augustin Cochin, restated Taine's ideas, with a more limited geographical application and more professional backing from the sources, in his *Les Sociétés de pensée et la révolution en Bretagne (1788–89)* (2 vols., Paris, 1925). He too attacked the men of the Enlightenment for taking refuge from the everyday world in 'an ideology of dreams' and for the unreality of their beliefs and ex-pectations. He added, however, to Taine's ideas an emphasis much heavier than any the master had provided on the role of active minorities in propagating the more extreme attitudes of the Enlighten-ment and more or less consciously destroying the foundations of the old regime. In this way, he argued, a kind of fictitious public opinion, divorced from reality, was built up.[1] What Cochin did, in fact, was to erect, on foundations laid by Taine, a type of 'conspiracy theory' of the fall of the old regime far more sophisticated and intellectually challenging than the crude allegations of this kind which had begun to circulate almost as soon as the revolution began. This view has remained very much a minority one. But his picture of the leaders of

---

[1] i. 15–23. See also 'Les Philosophes' in *Les Sociétés de pensée et la démocratie*, pp. 5–21. In the essays 'Comment furent élus les députés aux états généraux' and 'La Campagne electorale de 1789 en Bourgogne' in the same collection Cochin shows how the elections of 1789 were dominated by organized and some-times fanatical unrepresentative minorities. On Cochin generally see A. de Meaux, *Augustin Cochin et la genèse de la révolution* (Paris, 1928).

the Enlightenment as a sect, actively proselytizing, is now clearly vindicated. There was undoubtedly deliberate combination by them to monopolize positions of intellectual influence, to obtain appointments, pensions, and other fruits of power for their supporters and to victimize their opponents.[1] Taine and his followers therefore not merely mounted the heaviest of all rhetorical attacks on the Enlightenment in France but also exposed the selfishness and irresponsibility which often underlay it.

THE OLD HISTORIOGRAPHY AT ITS ZENITH

Taine was the last really important author to present a picture of the old regime in France which was dominated by personal feeling and unashamedly ideological. It is true that thirty years after *Les Origines de la France contemporaine* had begun to appear the great Socialist parliamentarian Jean Jaurès provided, in the first volume of his *Histoire socialiste de la révolution française*, a systematic and sustained Marxist view of the period.[2] But this was hardly an adequate left-wing response to the conservative fulminations of Taine. To Jaurès the old regime had collapsed because, led by a weak and indecisive monarchy too anxious to protect the privileges of a more and more functionless nobility, it could not retain the loyalty of a bourgeoisie increasingly conscious of itself as a class with interests and demands of its own. Against Taine he argued that the Revolution was rooted in realities, not the outcome merely of a destructive intellectual fashion or the malevolence of a small intellectual minority. It was the culmination of generations of commercial and industrial growth (Jaurès paid particular attention here to the development of the major French seaports during the last decades of the old regime and considerably exaggerated the real importance for the French economy of the expansion of colonial trade which

---

[1] For some interesting sidelights on this see R. Darnton, 'The High Enlightenment and the Low-Life of Literature in pre-revolutionary France', *Past and Present*, No. 51 (May, 1971).

[2] On Jaurès see F. Venturi, *Jean Jaurès e altri storici della rivoluzione francese* (Milan, 1948), pp. 13–109, reprinted in his *Historiens du XXᵉ siècle* (Geneva, 1966), pp. 5–66; H. Hirsch, 'Jean Jaurès als Historiker' in *Denker und Kämpfer. Gesammelte Beiträge zur Geschichte der Arbeiterbewegung* (Frankfurt, 1955), pp. 149–81; Madeleine Riberioux, 'Jean Jaurès, storico della rivoluzione francese', *Rivista storica del socialismo*, 1962, No. 17, pp. 591–611. The last gives a strongly Marxist view.

undoubtedly took place).[1] It was also the outcome of the increasing interest in eighteenth-century France in social and economic problems, an interest visible in the efforts to assemble reliable population statistics and in the considerable body of writing on the question of free trade in grain. Its intellectual background and inspiration were thus profoundly realistic; but Taine had completely ignored all this practical effort.[2]

The book presented the first sustained discussion of the old regime from a moderate and intelligent Marxist point of view. Its contribution to the historiography of the French Revolution was lasting and important. But it does not compare in originality with the work of de Tocqueville; and it lacks the distinctive personal flavour, of populist enthusiasm on the one hand, élitist pessimism on the other, which marks so strongly the work of Michelet and Taine. By the later nineteenth century conditions were becoming increasingly unfavourable to writing of this sharply individual and dogmatic kind. The collapse of the old regime was now a century distant. The Revolution was still the central event of French history, the one which above all still divided French society, not merely politically but into different intellectual and psychological worlds. Nevertheless, as it receded into the past its divisive power was inevitably though slowly dulled. Equally important, the writing of history was now rapidly becoming professionalized. The important books on the old regime produced down to the 1880s were the work of men who were not primarily teachers of history; Michelet is the only exception, and an incomplete one, to this generalization. In the later nineteenth century, however, we enter a new world. The growing popularity of history as a university subject, the resulting growth of a large body of professional teachers and researchers, the foundation of historical periodicals read only by professional historians, were rapidly changing this position. It was becoming increasingly rare (though as the example of Cochin proved not impossible) for a man who held no university post to make important large-scale contributions to

---

[1] See his discussion of the development of Bordeaux, Marseilles, and Nantes (*Histoire socialiste*, edn. revised by A. Mathiez (Paris, 1922), i. 68 ff., 73 ff., 78 ff.). The entire section on 'La Vie économique' (pp. 62–118) tends to overstress the sophistication of French economic life in the last years of the old regime. On the tendency of many historians to exaggerate the real importance of the growth of French colonial trade in the later eighteenth century see C.-E. Labrousse, *La Crise de l'économie française à la fin de l'ancien régime et au début de la révolution*, i (Paris, 1944), Introduction générale, p. xxxvii.

[2] *Histoire socialiste*, i. 50 ff.

historical knowledge. From the last decades of the century onwards, therefore, important writing on the old regime in France displays an increasing completeness and accuracy of information, an increasing sobriety of tone, an increasing tendency to deal with subjects limited chronologically or geographically to make them manageable under the new conditions, in a word an increasing professionalism.

The old type of account, encyclopedic and with a marked political bias, was not extinct. The republican and anticlerical Alphonse Jobez produced very large-scale examples of this kind of historical writing in his *La France sous Louis XV* (6 vols., Paris, 1864–73) and *La France sous Louis XVI* (3 vols., Paris, 1877–93). These enormous works (together they run to about 8,600 octavo pages) are a monument of industry. They contain a vast amount of usually accurate information and are based on a fairly wide range of materials— documentation from the Archives Nationales as well as memoirs, printed letters, and pamphlets. They are fuller, cooler, and more reliable than Michelet's treatment. But they are highly traditional works, conservative in their scope and form if not in the political ideas which rather obviously underlie them. They are throughout completely political, centred on Paris, the central government, and the court. The complex social and economic questions which were so much to exercise the writers of the twentieth century are almost completely outside the author's purview.[1] Even the administrative machine receives little attention, and local and provincial administration none at all. Both works are narratives, almost chronicles, conscientious, laborious, long-winded, and narrow. Even in their own day they must have seemed a little like museum pieces.

By the 1870s and 1880s a distinct tendency is visible towards writing of a more specialized kind than had been normal in the first three-quarters of the nineteenth century. Thus F. Rocquain, in his *L'Esprit révolutionnaire avant la révolution, 1715–1789* (Paris, 1878) produced the most complete and impartial study so far written of this aspect of the intellectual life of eighteenth-century France.

---

[1] He provides some account of the unrest among the silk-workers of Lyons in the 1730s and 1740s, and even prints as a *pièce justificative* the annual budget of a worker's family in 1744 (*La France sous Louis XV*, iii. 392–8, 475–7); but this is exceptional. More typical is the dismissal of the very important recoinage of 1726 in a couple of pages (ii. 419–21), a very meagre allowance of space in proportion to the immense length of the book. A fairly long discussion of administration and society in the French colonies in the early 1770s (*La France sous Louis XVI*, i. 381–99) contrasts forcibly with the lack of any discussion of similar problems in provincial France.

Though based entirely on well-known printed materials—the memoirs of Saint-Simon, Barbier, and Marais, the correspondence of Voltaire and Grimm—the book is even today not entirely without use. In particular it was probably the first to bring out the intensity of the interlinked crises of the early 1750s; there might well, Rocquain argued, have been a revolution in 1754 and again in 1771.[1] A few years later A. Chérest, in his *La Chute de l'ancien régime (1787–1789)* (2 vols., Paris, 1884) developed fully for the first time the idea, already stated by Lameth two generations earlier[2] and today a truism, that the fall of the old regime began, not with the meeting of the States-General or the fall of the Bastille, but with the summoning of the Assembly of Notables in February 1787. The book, though again based mainly on printed materials, was a thorough piece of work. It gave a considerable amount of attention to events in the provinces, though still much less than a present-day study of the same subject would. Almost simultaneously René Stourm, a former high-ranking civil servant, published in *Les Finances de l'ancien régime et de la révolution* (2 vols., Paris, 1885) the first really detailed and professional study of the tangled subject of old-regime finance. In 1879 E. Martin Saint-Leon produced the first professional study of trade guilds under the old regime in his *Histoire des corporations de métiers depuis leurs origines jusqu'à leur suppression en 1791;*[3] and in 1900–1 Émile Levasseur published in his *Histoire des classes ouvrières et de l'industrie en France avant 1789* the largest collection of information on the subject hitherto assembled (though the book is seriously limited, in the eyes of the present-day historian, by its dated outlook and its almost complete disregard of agriculture, overwhelmingly the most important industry in France during the eighteenth century). A decade later Camille Bloch provided the first serious professional study of official efforts to alleviate poverty in the last days of the old regime in his *L'Assistance et l'état en France à la veille de la révolution* (Paris, 1908). Almost two decades later again Gaston Martin published what is still perhaps the most balanced general study of Freemasonry and the Revolution, a subject which since the 1790s had excited passionate and ill-informed debate,

[1] Preface, p. viii, pp. 180–1.

[2] See p. 6 above.

[3] Though of this only a relatively small part deals with the eighteenth century. Of the 876 pp. of the third edition (Paris, 1922) which I have used, only pp. 505–617 cover this period.

in his *La Franc-Maçonnerie française et la préparation de la révolution* (Paris 1926).

Upon this body of increasingly competent research could be based by the 1930s and 1940s general works which summed up a great mass of specialized writing and which were the culmination of over two generations of professional effort. Such were *Les Origines intellectuelles de la révolution française* by Daniel Mornet (Paris, 1933) and P. Sagnac's *La Formation de la société française moderne* (2 vols., Paris, 1945–6). Both, especially the former, are remarkable works, already classics. Yet both books look back rather than forward; they summarize the work of the past rather than point to future developments. Mornet took a wide view of his subject and his book includes much precise information, not merely on intellectual history in a narrow sense but on the sales of different books during the eighteenth century, on the growth of the newspaper press in France, and on the *académies*, literary societies, libraries, and courses of lectures which were now stirring the hitherto stagnant intellectual life of the provinces.[1] Yet his view of French intellectual life is still overwhelmingly one from above, from the upper rungs of the social ladder. He was not ignorant of the fact that under the cultured and intellectualized surface there was an immense and almost impenetrable undergrowth of popular ignorance, conservatism and superstition.[2] But with all its great virtues and width of view the picture he paints is, by the standards of today, one too much dominated by traditional assumptions about the nature of his subject. Sagnac too saw his subject more from above, from the standpoint of Paris and the government, than would now be considered justified. His book contains relatively little on the life of the provinces (though in his bibliography he refers to some of the more important studies of this available when he wrote); and it is notable that the peasantry and its problems receive relatively little of his attention,[3] compared with that given to the nobility and middle classes. With all its achievements, writing of this kind had not said the last word. Even as Mornet and Sagnac were at work other historians were beginning a transformation, on several different levels, of our conception of old-regime France.

---

[1] 3rd edition (Paris, 1938), pp. 134–7, 159 ff., 299–318.

[2] See for example the illustrations he quotes of the survival of popular religious devotion, p. 216.

[3] 1st edition, pp. 73–4, 185–90, 240–4.

THE LAST FORTY YEARS: QUANTIFICATION, REGIONAL
STUDIES, MENTALITÉS

Three trends have dominated much of the work of the last four
decades, work which by its combination of thoroughness and imagination, its intellectual scope and use of illustrative detail, its sheer
professionalism, has made French historiography arguably the best
in the world. These are: large-scale efforts to use the great riches
of the archives of this period, particularly those of the royal administration, to quantify as many aspects as possible of eighteenth-
century French life (prices, demographic conditions, and even some
aspects of intellectual life); the production of a series of remarkable
studies of French provinces or geographical regions which attempt
to present a coherent picture of every aspect of their life; and more
recently a growing interest in popular culture, psychology, and ideas,
in those aspects of the intellectual, or at least non-material, life of the
French people which found no echo in the sophisticated world of the
enlightened minority. These three trends do not exhaust the new
departures made during recent decades; but it is in these directions
that the most fruitful and original work has been done and that a
sustained growth of knowledge has been most apparent.

The movement towards highly quantified studies of many aspects
of the old regime received its first great impetus from two works by
the economic historian C.-E. Labrousse, his *Esquisse du mouvement
des prix et des revenus en France au XVIIIe siècle* (2 vols., Paris, 1933),
and his *La Crise de l'économie française à la fin de l'ancien régime et
au début de la révolution*, i (all published, Paris, 1944). Labrousse
himself, however, was inspired by the work of the economist
François Simiand, whose *Le Salaire, l'évolution sociale et la monnaie:
essai de théorie expérimentale* appeared in 1932.[1] Simiand attempted,
like so many scholars before him, to make the social sciences
resemble more closely, in precision and certainty, the natural ones.
He did this by a study of wages in France during the period 1789–
1928 which, he believed, revealed long-term fluctuations, a Phase 'A'
of rise and a Phase 'B' of fall, the alternation of these phases being
determined mainly by changes in the supply of money. His work,
quite apart from the fact that it does not refer directly to the

---

[1] On Simiand in general see M. Lazard, *François Simiand* (Paris, 1936), and for
discussions of his ideas *Annales historiques de la révolution française*, ii (1930),
581 ff., iv (1932) 192 ff., v (1933), 161 ff., xiv (1937), 290–304; and J. Bouvier,
'Feu François Simiand', *Annales*, 28e année, No. 5 (Sept.–Oct. 1973), 1173–92.

eighteenth century, is relentlessly abstract and very difficult to read, and suffers from severe technical limitations. In particular it deals only with money and not with real wages, and only with very long-term fluctuations and not with the cyclical ones over shorter periods or seasonal ones within a single year which were of equal practical importance. Labrousse, unlike Simiand, wrote as a historian rather than a statistician or philosopher. His approach was more descriptive and more concrete; indeed he explicitly abandoned any attempt to explain the long-term price and wage movements which he discerned in France.[1] Moreover he was much more active than his master in the search for factual information with which to support his generalizations. He believed that three sorts of price movement could be discerned in old-regime France: firstly a movement 'de longue durée' extending over several decades; secondly cyclical movements of roughly ten years' duration delimited by years of crisis; and finally merely seasonal movements which depended primarily on weather conditions.[2] The eighteenth century in general was a period of rising prices in France; the country had enjoyed a long 'flux de prosperité' in 1733–1817. But this long-term rise was broken by pronounced falls covering in some cases periods of a good many years. Thus the prosperity of the old regime, at its apogee in 1763–70, was by the later 1770s endangered by a sharp decline in the prices of many essential commodities. The peasant producer of the cereals and above all wine[3] whose prices were falling found himself still paying a high rent based on the buoyant prices of earlier years and also raised by the competition for land which resulted from the slow but steady growth of population. Rents and seigneurial dues, as a proportion of the agricultural producer's total income, thus became much more onerous as prices fell;[4] and there was a tendency for resources to be diverted from agriculture and the countryside to the towns, luxury consumption, and industry. However the textile industries, dominant in this sector, were stagnant during the 1770s

---

[1] *Esquisse*, i. Introduction, xxiv–xxv.

[2] Ibid. xxiii–xxiv. Later, in *La Crise*, Introduction générale, ix–xi he refined and complicated this picture by discovering fluctuations which covered one or two years without any fixed periodicity and intercyclical ones which covered one ten-year cycle and part of another.

[3] Part II of *La Crise*, about two-thirds of the book, is devoted to a very detailed study of the collapse of income and profits in wine-growing in the years 1778–91.

[4] In the period between 1730–9 and 1770–90 rents, he concluded, rose by 82 per cent while the price of wheat increased by only 56 per cent (*Esquisse*, ii. 379–80).

and 1780s; and there was severe industrial unemployment in the years before the Revolution. Money wages in general had risen by less than a quarter between 1726–41 and 1785–9; of the sixty-one wage series he was able to construct for the period the rise in almost half was of less than 11 per cent.[1]

In essence, Labrousse achieved two things. He assembled, principally in the *Esquisse*, a vast amount of information on price and wage movements in eighteenth-century France; and he provided, above all in *La Crise*, the most sophisticated and elaborate discussion hitherto produced of the extent and nature of economic strain in the last years of the old regime. The masses, it could now be argued more plausibly than ever before, had been led into sympathy with the Revolution not by their own inherent depravity, as Taine had assumed, or by a conspiracy of the 'enlightened', as Cochin had argued, but by measurable economic pressures.[2]

It is not difficult to criticize some of the work of Labrousse on technical grounds as over-simplified and over-mechanistic.[3] Some of his statistical treatment of the material he so assiduously collected could now well be regarded as rather narrowly commonsensical and unsophisticated.[4] It is also likely that the figures which he provides in such profusion sometimes convey, as so often when the historian deals with a pre-statistical age such as that of the old regime, a misleading impression of precision. Nor was his work entirely new, since at least one effort of the same kind, though on a very much smaller scale, had been made towards the end of the nineteenth century.[5]

[1] Ibid. 491.

[2] It should be pointed out, however, that Labrousse somewhat modified, between the publication dates of the two books, his ideas on the extent to which the Revolution was caused by economic factors. In 1933 he wrote that 'the economic situation created in great part the revolutionary one' (*Esquisse*, ii. 640); in 1944 he argued that 'an immense error of imputation makes the political crisis arise from an economic one' (*La Crise*, Introduction générale, p. xlviii). There is a convenient study of the work of Labrousse from this standpoint in G. Lefèbvre, 'Le Mouvement des prix et les origines de la révolution française', *Annales historiques de la révolution française*, xiv (1937), 289–329, reprinted in Lefèbvre's *Études sur la révolution française* (Paris, 1954).

[3] See the interesting comments of D. Landes, 'The Statistical Study of French Crises', *Journal of Economic History*, x (1950), 195–211.

[4] For example the discussions of price-averaging and of means by which the reliability of his figures may be tested in *La Crise*, i. 130 ff., 136 ff.

[5] D. Zolla, 'Les Variations du revenu et du prix des terres en France aux XVIIe et XVIIIe siècles', *Annales de l'école des sciences politiques*, 1893–4. The *Histoire économique de la propriété, des salaires, des denrées et de tous les prix en*

But whatever its weaknesses his work gave a powerful impetus to the movement for greater precision and more quantitative treatment in economic and social history which was already beginning to take shape by the end of the 1920s. Several years before Labrousse published his first book Georges Lefèbvre had brought out his *Les Paysans du Nord pendant la révolution* (Lille, 1924: reissued Bari, 1959), one of the classical works of revolutionary historiography. The first part of this provided a very detailed and to a considerable degree quantified study, based on the use of well over a thousand archive collections, of the rural life of the Nord department in the last decades of the old regime; it can perhaps be regarded as the first serious attempt to write quantified social history. In 1929 the *Annales d'histoire économique et sociale* was founded in Paris with the object of widening the historian's view of his subject and of fusing together the different aspects of it—economic, social, political, and administrative—which hitherto had usually been treated in isolation. An International Scientific Committee on the History of Prices was set up shortly afterwards under the chairmanship of Sir William Beveridge: under its auspices Henri Hauser published in 1936 his not entirely satisfactory *Recherches et documents sur l'histoire des prix en France de 1500 à 1800*.

In the 1930s this quantifying effort was concentrated above all on price history; and much effort has continued to be devoted to the improvement and refinement of price series.[1] After the second World War, however, the drive towards quantification increasingly invaded many other aspects of the subject. Demography was by its very nature the most obvious of all candidates for its attentions; and in 1946 Jean Meuvret published a pathfinding article in which for the first time he explored in a rigorous manner the connection in France between high food prices and abnormally high mortality, as in the terrible years of 1693 and 1709.[2] He showed how the intensity of such crises could be measured by the yardstick of the increasing ratio during them of deaths to conceptions: he also demonstrated

---

général, depuis l'an 1200 jusqu'en l'an 1800 of the Vicomte G. d'Avenel (7 Vols., Paris 1894–1926) belongs to a different and less rigorous tradition.

[1] See the methodological discussions in P. Vilar, 'Histoire des prix: histoire générale', *Annales*, 4ᵉ année (1949), No. 1, 29–45; and 'Remarques sur l'histoire des prix', *Annales*, 16ᵉ année (1961), No. 1, 110–15.

[2] J. Meuvret, 'Les Crises de subsistances et la démographie de la France d'ancien régime', *Population*, 1946, pp. 643–50.

that they became much less acute during the second half of the eighteenth century—the most convincing tribute hitherto paid by any historian to the achievements of the old regime in its last decades. During the last generation a school of historical demographers, led by Louis Henry of the Institut national d'études démographiques has transformed the study of the subject and set standards which the rest of the historical world must struggle to equal. In particular the richness of the raw materials available in France for this type of work has allowed the technique of 'family reconstitution'[1] to be developed into a tool with which the historian can bring himself face to face with the demographic (and therefore to a large extent social) realities of past ages. The output of work of this kind, above all in the form of articles, is too great to be detailed here;[2] and in recent years the use of computerized methods has opened new perspectives by greatly easing the task, inherent in this type of work, of collecting and correlating vast amounts of minutely detailed information.

Perhaps more important, and usually easier for the non-mathematical historian to understand, are the efforts which have been made in recent years to apply strictly quantitative methods and criteria to the study of different social groups in the France of the old regime. An outstanding example of this is the *Structures et relations sociales à Paris au XVIIIe siècle* of Adeline Daumard and F. Furet (Paris, 1961: Cahiers des annales, 18). This is based on a detailed analysis of 2,597 marriage-contracts signed in Paris in 1749; these are used as a guide to the professions, wealth, social relationships, and ambitions of a large part of the population of the city. Inevitably there are limits to what can be achieved by such methods. Their value when applied to eighteenth-century society is not yet beyond question. Though about sixty per cent of marriages in Paris

---

[1] The reconstruction of families normally on the basis of the mentioning of their individual members in parish registers and similar documents. The technique was first suggested by Henry in his 'Une richesse démographique en friche: les registres paroissiaux', *Population*, 1953, No. 2, pp. 281–90. It was first put into practice on a large scale in E. Gautier and L. Henry, *La Population de Crulai, paroisse normande: étude historique* (Paris, 1958) which is concerned with a village of about 1,000 people and is based on the reconstitution of families for the period 1674-1742.

[2] See, to give only one example, *Annales*, 27e année, Nos. 4–5 (1972) for a series of the most recent articles, often highly technical, on family reconstitution and other aspects of historical demography. For an up-to-date view of the subject see A. Santini, 'Tecniques [*sic*] and methods in historical demography (17th–18th centuries)', *Journal of European Economic History*, i. No. 2 (Fall 1972), 459–69.

gave rise to marriage-contracts the really poor, who married without such legal provision, are excluded from the scope of such a survey. Much more serious, it can be argued that such an approach puts too exclusive an emphasis on wealth as a determinant of social status, and thus plays down unduly the ideas of *dignité* and *honneur* (as represented, for example, by military rank and service to the monarch generally) which played so great a part in determining rank in eighteenth-century society. France before the revolution, it has been argued, was a closed society of orders and groups. Labrousse and his followers, by treating it as though it were an open one of the type generated by the Industrial Revolution, and by dividing the population into socio–professional groups of a kind appropriate only to the nineteenth and twentieth centuries, have confused and over-simplified the real position.[1] Nevertheless quantitative studies may allow the historian to confront conventional and somewhat vague ideas with the usually much more complex realities of a social structure. In particular the Daumard–Furet analysis throws much light on social mobility as evidenced by choice of marriage-partners and of witnesses to the contracts. It seems to open the way to a series of similar studies of other social groups.[2]

An even more ambitious effort at the quantified study of society is M. Couturier, *Recherches sur les structures sociales de Châteaudun, 1525-1789* (Paris, 1969), the first work to attempt a computerized treatment of such problems.[3] The statistics of the effort involved are impressive: 200,00 punched cards were used, and about 1,200,000 perforations, needing forty days' work by a punch-card operator, were made in them. The potentialities for the future are equally striking: the author argues that computers could well be used to construct from such material as tax-returns, census-returns, or notarial archives the biographies of large numbers of individuals.

---

[1] See the powerful criticisms of R. Mousnier, 'Problèmes de méthode dans l'étude des structures sociales des XVIe, XVIIe, XVIIIe siècles', reprinted in his *La Plume, la faucille et le marteau: institutions et société en France du moyen age à la révolution* (Paris, 1970), pp. 14 ff.

[2] See the comments of R. Mandrou, *La France aux XVIIe et XVIIIe siècles* (Paris, 1967) pp. 255–8. Furet himself has discussed the possibilities of using the 'taxe des pauvres', which was graduated according to the wealth of the taxpayer, as a guide to the social structure and distribution of wealth in Paris in the 1740s (*Annales*, 16e année (1961), No. 5, 939–58).

[3] There is a long discussion of this book in A. Menzione, 'Storia sociale quantitativa: alcuni problemi della richerca per i secoli XVI–XVIII', *Studi storici*, xii (1971), 585–96.

Collections of such biographies could be built up and would then provide an unprecedentedly secure and complete basis for the study of different social groups.[1] Clearly, moreover, to use quantitative methods is not necessarily to dehumanize social history, as critics of the new approaches are fond of alleging. In Part IV of his book M. Couturier is able to reconstruct in great detail a single street (the Rue Saint-Lubin) in eighteenth-century Châteaudun and almost every aspect of the lives of its inhabitants. This achievement (made possible largely by the existence of a list of 1743 giving the names of the inhabitants and the positions of their houses, which acts in turn as a lead to the very numerous other documents available) makes most of the social history written during the nineteenth and even the early twentieth centuries look superficial by comparison. Sometimes, it is true, quantification becomes an end in itself and the amassing of details outruns the author's ability or even willingness to draw useful conclusions from them.[2] The full potentialities and exact limitations of this type of study are still not clear, especially as the application of computer techniques to historical research is still in its infancy. But it seems clear that though the new methods may not change fundamentally many of our ideas about the past they can and will add a new fullness and precision to our knowledge of many aspects of it.

Quantification has assumed other and less obvious forms. The history of climate, for example, so vital for the harvest and therefore in pre-industrial societies sometimes literally a matter of life and death, offers surprising possibilities for work of this kind. The movements of glaciers and the dates of the beginning of the wine-harvest (for both of which there is relatively reliable information) as well as the records of temperature and precipitation which began to be kept from about 1700 onwards in many parts of western Europe, have been used for this purpose by Emmanuel Le Roy Ladurie, one of the most brilliant and imaginative of contemporary French historians.[3]

[1] There is a long discussion of the methods employed and of their difficulties and potentialities in Part I, chap. i of the book. See also the long methodological introduction in Régine Robin, *Société française en 1789: Semur en Auxois* Paris, 1970).

[2] The somewhat disappointing M. El Kordi, *Bayeux au XVIIe et XVIIIe siècles* (Paris–The Hague, 1970) is perhaps an example of this.

[3] See his 'Climat et récoltes aux XVIIe et XVIIIe siècles', *Annales*, 15e année (1960), No. 3, 434–65; and on a larger canvas his *Histoire du climat depuis l'an mille* (Paris, 1967).

Perhaps the most impressive single example of all of the use of quantitative techniques is the great book of André Corvisier, *L'Armée française de la fin du XVIIe siècle au minstère de Choiseul: Le Soldat* (Paris, 1964). This huge volume of almost 1,100 pages, with its sixty-seven maps and graphs, its fourteen appendices, its four different indices, its enormous bibliography, is one of the modern masterpieces of professionalism in the writing of history. Moreover it is based on the strictest quantitative methods; over 85,000 individual soldiers were studied (mainly in the form of three large samples taken from the years 1715, 1737, and 1763) and data-processing methods involving the use of 50,000 punched cards were employed. (This analysis, when it was begun in 1954, was almost certainly the earliest use of such methods in the study of any aspect of the old regime in France.) Even with such mechanical aids there were certain lines of inquiry—losses in wars and through epidemics, the influence of economic fluctuations on desertion—which Corvisier could not undertake because of the immense labour involved and the uncertainty of achieving any useful result.[1] Nevertheless he has been able to give a new precision and certainty to some of the assertions about French military life in the eighteenth century which had been put forward by earlier writers, and also to make important discoveries of his own. He has shown conclusively, for example, that the poverty or wealth of a region had little to do with the number of recruits it produced for the French army; it was the frontier areas in the north and east, with a strong military tradition, others on the main lines of military communication and some with a tradition of emigration, such as Alsace, which yielded relatively large numbers of recruits, not the poor provinces of the south and west.[2] He has even demonstrated the growing effectiveness of discipline in the French army by the fact that from the end of the War of the Spanish Succession onwards the wounds received in battle by soldiers tended more and more to be on the left side—the one most exposed to the enemy.[3] Above all the book presents the army as a part of society; the stress throughout is on human realities rather than military techniques or military organization for its own sake.

The most ambitious and intellectually sophisticated effort at the use of quantification in the study of eighteenth-century France

[1] See the general description of his methods, Introduction, pp. x-xii.
[2] See the summary of his conclusions on pp. 967–8.
[3] p. 978.

came in 1961 with an effort, associated with the Institut de science économique appliquée, to produce a great quantitative history of the French economy. It failed. The attempt to unite economists and historians in a joint enterprise soon broke down and has not been repeated. There have been in particular complaints from some French historians that their economist colleagues regard them as mere providers of raw data upon which a structure of 'retrospective econometrics' can be erected, and claims that the historian should look for more refined methods of classification and a more precise vocabulary rather than for the general laws and statistical verifications of them sought by the economist.[1] Moreover the most uncompromising effort to write the history of the French economy in the eighteenth century along strictly quantitative lines, J.-C. Toutain's *Le Produit de l'agriculture française de 1700 à 1958*, vol. i, *Estimation du produit au XVIIIᵉ siècle* (Paris, 1961), shows well the great complexities of the task. In particular there is no equivalent so far as agricultural production is concerned to the very abundant materials relating to prices on which Labrousse was able to base his work, while great regional differences in productivity are another major source of difficulty. The result is that the three separate estimates which Toutain was able to produce differ widely and clearly contain wide margins of error. A more recent writer on the same subject has been reduced to working backwards from the figures yielded by the statistical survey of French agriculture carried out in 1840, the earliest which is generally reliable.[2]

Quantification in the writing of history has obvious limitations. Quite apart from the fact that in a pre-statistical age the figures available to the historian will very often, perhaps normally, be inaccurate or incomplete, they can never take the place of verbal

[1] See the discussion between the economic historian Pierre Chaunu and the economist Jean Marczewski, the scholar most closely associated with the new enterprise, printed in *Cahiers Vilfredo Pareto* (Geneva), 1964, No. 3, 125–80; also the comments of Pierre Vilar, 'Pour une meilleure compréhension entre économistes et historiens: "Histoire quantitative" ou "Econométrie retrospective"', *Revue Historique*, ccxxxiii (1965), 293–312. The hopes and plans which underlay the project are outlined in R. Mandrou, 'Pour une économie historique: le revenu national français de 1726 à nos jours', *Annales*, 15ᵉ année, (1960), No. 3, 752–8.

[2] M. Morineau, 'Y a-t-il eu une révolution agricole au XVIIIᵉ siècle?', *Revue historique*, ccxxxix (1968), 299–326. His conclusions have recently been elaborated and amplified in a book, *Les Faux-semblants d'un démarrage économique: agriculture et démographie en France au XVIIIᵉ siècle* (Paris, 1971).

definition and analysis. They are, after all, assembled and grouped in answer to questions expressed in words; they can never replace definitions and concepts. Though they can measure wealth or poverty they are much less able to tell us how and why individuals or groups became rich or poor.[1] Nevertheless the quantitative approach has done more than anything else to justify, at least in France, the claim that 'the vitality of economic and social history has no equal in the historical world at the moment'.[2]

Regionalism and provincial patriotism were in the nineteenth century powerful forces in much French historical writing. The provincial and regional histories which had already begun to appear under the old regime (their production almost certainly stimulated by the growing interest from about 1750 onwards in the rights of estates and other provincial institutions) founded a tradition which by the beginning of the twentieth century was producing professional work of high quality. Until fairly recently, however, this type of writing has remained essentially traditional. Much of the considerable output of it during the nineteenth century was legal or antiquarian in its outlook.[3] Even in the first decades of the twentieth century it usually lacked both the interest in quantifiable factors such as prices, wages, and demographic realities, and that in popular culture and the less sophisticated aspects of intellectual life, which are characteristic of so much of the best recent French historical writing on the eighteenth century. Brittany, which has attracted more attention than any other French province, is a good case in point. Henri Sée's *Les Classes rurales en Bretagne du XVIe siècle à la révolution* (Paris, 1906) is a large book based on extensive research in archives; it is clearly the work of a first-class professional historian. Yet it completely lacks the quantitative treatment which a present-day writer would undoubtedly give to the same subject. It contains by

---

[1] See the caveats in A. Soboul, *La Civilisation et la révolution française:* vol. i, *La Crise de l'ancien régime* (Paris, 1970), pp. 32–3.

[2] Mandrou, *La France aux XVIIe et XVIIIe siècles*, p. 270.

[3] e.g. C. J. Trouvé, *Essai historique sur les états généraux de la province de Languedoc* (2 vols., Paris, 1818); G. Bascle de Lagrèze, *Histoire du droit dans les Pyrenées—comté de Bigorre* (Paris, 1867) and *La Société et les mœurs en Béarn* (Pau, 1886); D. Mathieu, *L'Ancien Régime dans la province de Lorraine et Barrois, d'après des documents inédits, 1698–1789* (Paris, 1879); M. Cohendy, *Mémoire historique sur les modes successifs de l'administration dans la province d'Auvergne . . . depuis la féodalité jusqu'à la création des préfectures en l'an VIII (1800)* (Clermont-Ferrand, 1856).

modern standards astonishingly little on prices[1] and no discussion at all of demography. Similar gaps, by present-day standards, can be found in the other, often excellent, examples of provincial and regional history produced in the early twentieth century.[2] A Rébillon, *Les États de Bretagne de 1661 à 1789* (Paris–Rennes, 1932) is a substantial administrative study, thorough and professional. But it makes no effort to explore the relationship between the estates and Breton society or anything but the official aspects of the subject such as their organization, powers, and finances. Much the same is true of the large work of H. Fréville, *L'Intendance de Bretagne (1689–1790)* (3 vols., Rennes, 1953) which is a straightforward series of studies, chronologically arranged, of the intendants of this period and of the way in which each administered the province.

By the 1930s there was a growing demand from a group of French scholars (which included geographers, economists, sociologists, and experts in liguistics as well as historians) for a more all-embracing type of historical writing. This would attempt to use the insights and resources of a wide range of disciplines to present so far as possible a complete and integrated picture of every aspect of human life, preferably within the bounds of some naturally-defined geographical area. Such a demand involved a movement away from the more official and political aspects of history towards the social, the economic, even the psychological ones, and almost always a heavy emphasis on the interrelationship of man and his physical environment. Lucien Febvre, the historian who more than anyone else typified the new approach, demanded 'history which smells of the good earth, of the countryside, of toil and the harvest'.[3] The *Annales d'histoire sociale* (which became in 1946 *Annales: économies, sociétés, civilisations*) was founded to propagate these ideas and has been for a generation or more the most intellectually adventurous historical periodical in the world. Nevertheless, although the 1940s

---

[1] Appendix I prints an official list of grain prices in 1733 and on pp. 411–13 the point is made, with a few supporting figures, that they could fluctuate sharply; otherwise the subject is almost ignored. H. Frotier de la Messelière, *La Noblesse en Bretagne avant 1789* (Rennes, 1902) is interesting on the economic position of the Breton nobility, but in a traditional and unquantitative way. Of its 961 pages only twelve (pp. 55–66) deal with population questions and only thirteen (pp. 145–57) with harvest fluctuations.

[2] L. Dutil, *L'État économique du Languedoc à la fin de l'ancien régime* (Paris, 1911) is a large-scale example.

[3] The phrase occurs in his *Combats pour l'histoire* (Paris, 1953), p. 393. See also in general his *La Terre et l'évolution humaine* (Paris, 1922).

and 1950s produced excellent writing on the history of French provinces and regions during the eighteenth century, some of it influenced by the increasing interest in quantification which has been mentioned above,[1] it was not until 1960 that Pierre Goubert published, in his *Beauvais et le Beauvaisis de 1600 à 1730: contribution à l'histoire sociale de la France du XVIIᵉ siècle* (2 vols., Paris, 1960), the first truly great work of the type which Febvre had envisaged. Though this book covers only the first decades of the eighteenth century it throws much light on that period as a whole, in particular by its careful underlining of the contrast, first drawn by Meuvret in 1946, between the terrible demographic crises of 1693–4 and 1709 and the much milder food shortages, the 'disettes larvées' of the eighteenth century proper.[2] The book contains no discussion, at least directly, of the intellectual life or popular culture, the *mentalités*, of the area. But to writers of economic and social history Goubert's work remains a model which they must strive to emulate but which has never been surpassed. It combines with striking effect the interest in human and environmental realities, which stems above all from Febvre, with that in economic fluctuations and crises, quantitatively studied, which is the legacy of Simiand and Labrousse. It is from this combination that a whole series of great regional studies published during the last decade which can be no more than mentioned here— Saint-Jacob on Burgundy, Baehrel and Agulhon on Provence, Rascol on the Albigeois, Poitrineau on Auvergne, Le Roy Ladurie on Languedoc, Meyer on Brittany—has been nourished and inspired.[3] Side by side with these huge and intimidating studies, with their buttressing of maps, graphs, statistical tables, appendices and bibliographies, stand some very detailed monographs of a more purely economic kind, notably the books of Pierre Dardel on the trade of

---

[1] The treatment of food prices and demography in D. Ligou, *Montauban à la fin de l'ancien régime et aux débuts de la révolution, 1787-1794* (Paris, 1958), pp. 163–88 is a good example.

[2] The discussion of the demography of the Beauvais area in Pt. I, chap. iii is a model of such work: see also the discussion of demographic fluctuations in Pt. II, chap. viii.

[3] P. de Saint-Jacob, *Les Paysans de la Bourgogne du Nord au dernier siècle de l'ancien régime* (Paris, 1960); R. Baehrel, *Une Croissance: la Basse-Provence rurale (fin XVIᵉ siècle –1789). Essai d'économie historique statistique* (Paris, 1961); M. Agulhon, *La Vie sociale en Provence intérieure au lendemain de la révolution* (Paris, 1970); A. Poitrineau, *La Vie rurale en Basse Auvergne au XVIIIᵉ siècle (1726–1789)* (2 vols., Paris, 1965); E. Le Roy Ladurie, *Les Paysans de Languedoc* (2 vols., Paris, 1966); J. Meyer, *La Noblesse bretonne au XVIIIᵉ siècle* (2 vols., Paris, 1966).

Rouen and Havre,[1] a very detailed study of a feudal seigneurie,[2] and at least one interesting and original work in English by an American sociologist.[3] It is above all on this mass of erudition, these serried ranks of facts and ideas, that modern general works on the history of eighteenth-century France are based,[4] and to them that they owe their excellence.

The movement towards quantification and the new large-scale regional studies both reflect, in different ways, a tendency towards concentration on social, economic, and institutional structures and away from the history of ideas in the conventional sense of that term. From the later eighteenth century onwards it had never been doubted that the movement of ideas, the influence and ramifications of the Enlightenment, was of fundamental importance in the last decades of the old regime. The study of ideas in this sense, however, was by its very nature a study of small intellectual élites. During the last generation it has been increasingly supplemented, if not superseded, by attempts to analyse the states of mind and unspoken assumptions, the instinctive beliefs and prejudices, in a word the *mentalités*, of unsophisticated and often uneducated people. This change of focus from the élite to the mass, from Paris to the provinces, probably springs from a number of sources. On the one hand it seemed by the 1930s that Mornet in his *Les Origines intellectuelles de la révolution française* and on a wider scale Paul Hazard in his *La Crise de la conscience européenne, 1680–1715* (3 vols., Paris, 1934) and *La Pensée européenne au XVIIIᵉ siècle: de Montesquieu à Lessing* (3 vols., Paris, 1946) had largely exhausted the traditional aspects of the subject. On the other, the growing influence of Marxism in most aspects of French intellectual life made studies of the 'masses' seem

---

[1] P. Dardel, *Navires et marchandises dans les ports de Rouen et du Havre au XVIIIᵉ siècle* (Paris, 1963), and *Commerce, industrie et navigation à Rouen et au Havre au XVIIIᵉ siècle* (Rouen, 1966).

[2] A. Pleisse, *La Baronnie de Neubourg: essai d'histoire agraire, économique et sociale* (Paris, 1961).

[3] C. Tilly, *The Vendée* (London, 1964).

[4] R. Mandrou, *La France aux XVIIᵉ et XVIIIᵉ siècles* (Paris, 1967); A. Soboul, *La Civilisation et la révolution française:* vol. i, *La Crise de l'ancien régime* (Paris, 1970); E. Labrousse, P. Leon, and others, *Histoire économique et sociale de la France*, vol., ii, *Des derniers temps de l'âge seigneuriale aux préludes de l'âge industriel (1660–1789)* (Paris, 1970). The last contains, in the Introduction by Labrousse, one of the most uncompromising assertions by any historian of the primacy of statistical series over all other types of raw material for the writing of economic history (pp. ix–xi); and it is significant that Part I of the book, by Goubert, deals entirely with demography.

more attractive and rewarding than ever before. Moreover the coming of the new quantitative techniques held out the opportunity of applying them in a field so far little touched by methods of this kind.

The achievements of this new attitude to the intellectual (perhaps it would be better to say non-material) life of eighteenth-century France are difficult to evaluate with any accuracy. Statistical and quantitative techniques are not easy to apply to mass psychology with reliable results, especially when the raw materials for such a study are incomplete and difficult to evaluate.[1] In the study of religious belief, for example, a field in which this type of approach has been widely used in the last two decades, it is doubtful whether the insights of Bernard Groethuysen, in his *Les Origines de l'esprit bourgeois en France* (Paris, 1927: English translation, London, 1968) have been superseded or even much altered by the work done since, though he used completely traditional literary and non-quantitative methods. As a sociologist rather than a historian he illuminated with great skill the cleavage which existed by the eighteenth century between the traditional, unintellectual, life-pervading form of religion still cherished by the masses and the more self-conscious, sceptical, and intellectualised beliefs of the élite. The work of the last generation has, however, shown in detail how sharp this distinction was, how deeply traditional and dominated by superstition and ritual the world-view of the mass of Frenchmen remained during a period of great intellectual activity among the upper ranks of the intellectual hierarchy. It has also brought out very clearly, in a number of excellent regional studies, the size of geographical variations in religious fervour and piety—the extent to which the large towns and the areas most affected by them were generally not now highly religious, whereas backward and traditional areas such as Brittany and the Auvergne still were.

The largest and most important of these studies, originally a thesis in demography (again the pervasive quantitative element), is the impressive work of F. Lebrun, *Les Hommes et la mort en Anjou aux 17ᵉ et 18ᵉ siècles: Essai de démographie et psychologie historiques* (Paris–The Hague, 1971). Part iii of this book gives a remarkable account of the survival of traditional beliefs and attitudes in a back-

---

[1] For a discussion of some of the problems involved see R. Mols, 'Emploi et valeur des statistiques en histoire réligieuse', *Nouvelle revue théologique*, 1964, pp. 388–410.

ward and economically stagnant part of France. Lebrun points out, for example, that as medicine became more scientific and discarded its magical aspects it lost the confidence and trust of ordinary people who still expected and demanded that it be mysterious and semi-magical.[1] Belief in the curative activities of saints and relics associated with them; the very free use in astrological and magically curative formulas of Christian terminology and the sign of the cross; the horror aroused by suicide and by the allowing of a child to die without baptism;[2] the widespread belief in traditional signs of approaching death (the appearance of bats or certain birds, the howling of dogs); the continuing enormous importance in preaching and popular religious feeling of the Last Judgement and the pains of Hell—all these paint a striking picture of a population quite untouched by the Enlightenment and still in many respects almost savages. On the other hand Lebrun shows that from about 1760 this position was slowly changing. The ills of the flesh were beginning to be thought of not as the irresistible wrath of God but as something which could be fought; 'the desacralization of disease and death' was under way. There was a marked disappearance of religious formulas from wills and the first moves towards the secularization of charity were beginning.[3]

Work of the type which Lebrun's book typifies has not ousted religious history of a more traditional kind. There has continued, for example, to be a considerable amount of writing on the well-worn but still unexhausted theme of Jansenism and its political and social significance in France during the eighteenth century.[4] But the new approach to the subject has changed and widened it, making it more human, more sociological,[5] less simply intellectual and institutional. It has also emphasized heavily the importance for the

[1] pp. 394–5.

[2] He quotes the striking case in 1718 of a young pregnant girl whose body, after she had killed herself, was exhumed and dragged face downwards through her town. In the *place publique* the executioner extracted the foetus from her body; it was then buried in the churchyard where unbaptized children were usually interred. The mother's body was hanged for an hour by the feet, bearing a written denunciation of her crime, then burnt and the ashes scattered (p. 422).

[3] pp. 435, 452–3. This consciousness of change over time is a marked superiority of recent work of this type to the book of Groethuysen, which treats the eighteenth century as a unity and shows no realization at all of development within it.

[4] e.g. R. Taveneaux, *Jansenisme en Lorraine, 1640–1789* (Paris, 1960) and *Jansenisme et politique* (Paris, 1965).

[5] e.g. L. Percuas, *Le diocèse de La Rochelle de 1648 à 1724. Sociologie et pastorale* (Paris, 1964).

subject of a range of sources hitherto unused by historians. One of the most original recent contributions in this area has attempted, for example, to study the ordinary man's ideas about death and the hereafter through an analysis of eighteenth-century sculptured representations of souls in Purgatory.[1] There is certainly need for an expansion and systematization of studies of popular religion, of the veneration of saints and relics, of parachristian beliefs of many sorts; and at the Sorbonne the seminar of Professor Dupront began in the mid-1960s a collective study of the history of pilgrimages in western Europe.[2]

Apart from religion popular reading and reading-habits are the aspect of French life in the eighteenth century which has attracted most attention from practitioners of the new cultural history. Here, as in the field of religion, there has been a swing of interest away from the educated and sophisticated to the common man, poor, semi-literate and barely touching the fringes of intellectual life. Here again quantification has played a considerable role. In particular the almanacs, popular romances, and booklets of folk-wisdom, which made up most of what was read by the French peasant or artisan and which were distributed by pedlars and beneath the notice of booksellers, have been studied in detail, notably by Geneviève Bollême. Both the main collection of literature of this kind, the Bibliothèque bleue (so called from the colour of the cheap paper in which the booklets were bound) and the almanacs, have been the subject of books by her.[3] This type of reading matter had already attracted a certain amount of attention in the nineteenth century;[4] but much of this was antiquarian or bibliographical rather than strictly historical. Now it is possible to see more clearly than ever before both the size of this literary *demi-monde* (there were in the eighteenth century 150 or more printers producing booklets of this

---

[1] Gaby and M. Vovelle, *Vision de la mort et de l'au-delà en Provence d'après les autels des âmes du purgatoire* (Paris, 1970: Cahiers des annales, 29).

[2] R. Mandrou, *La France aux XVIIᵉ et au XVIIIᵉ siècles*, pp. 276 ff.

[3] *La Bibliothèque bleue: la littérature populaire en France du XVIᵉ au XIXᵉ siècle* (Paris, 1971); *Les Almanachs populaires aux XVII et XVIII siècles: Essai d'histoire sociale* (Paris–The Hague, 1969). See also her article on 'Littérature populaire et littérature de colportage au 18ᵉ siècle', in G. Bollême and others (eds.), *Livre et société dans la France du XVIIIᵉ siècle* (Paris–The Hague, 1965).

[4] Champfleury (=J. F. F. Husson), *De la littérature populaire en France. Recherches sur les origines de la légende du Bonhomme Misère* (Paris, 1861); C. Nisard, *Histoire des livrets ou de la littérature de colportage* (Paris, 1854); J. Grand-Carteret, *Les Almanachs français, 1600–1895* (Paris, 1896).

kind in about seventy different places, mostly in northern France), and its highly traditional nature (one of its most popular products, for example, *L'Escole de Salerne*, a poem giving general advice on health, went through 300 editions between 1474, when it first appeared, and 1846). It is also, more interestingly, possible to detect in the almanacs the advent in the second half of the century of a more modern world picture. They become objective, moderate, rational, and even sceptical to an extent unknown in the past. The attitude to death which they embody becomes more matter-of-fact; they place more stress on the importance of feeling, of the good qualities of the heart; the position of women as reflected in them improves; and they display an increasing interest in public affairs.[1] Nevertheless it is clear that this varied literature (as well as almanacs it includes lives of saints, fairy-tales, short novels, songs, burlesques, and short books on education and games) was above all a literature of escape. It was called into existence by the need of hard-worked people leading poor and insecure lives for some vision of a world different from the difficult and often tragic one which they inhabited. This literature of the common man has as its most remarkable (though also its quantitatively smallest) component a group of legendary historical tales (notably about Charlemagne and his paladins) with a strong magical element; in it the common folk never appear as actors.

The work of Madame Bollême contains a substantial element of quantitative analysis. This is also visible on a wider scale in the efforts which have been made in recent years to list and analyse by subject the entire output of the printing press in eighteenth-century France. As long ago as 1912 Mornet began pioneering but incomplete efforts in this direction. These have now been carried to a much higher pitch of completeness and professionalism. In particular R. Estivals has produced, in his *La Statistique bibliographique de la France sous la monarchie au XVIIIᵉ siècle* (Paris–The Hague, 1965) a very ambitious and complete effort of this kind which is, significantly enough, dedicated to Labrousse. With all its virtues it also displays the tendency towards abstraction and dogmatism[2] which mars a good deal of historical writing of this essentially quantitative

[1] See the summary in Bollême, *Les Almanachs*, pp. 110–13, based on a reading of about 500 examples.

[2] For example in the rather unconvincing effort (pp. 410–12) to show that fluctuations in French book production from the later seventeenth century onwards fit in well with the system of economic fluctuations proposed by Simiand.

type. There have been several attempts to show how changes in the subject-matter of the books published in this period can be used as a guide to a changing intellectual climate. Thus François Furet has demonstrated in a strictly quantitative manner how the share of theology in French book production declined, especially during the second half of the century, while that of the sciences and arts, and to a slightly less extent of history, rose. The share of poetry also fell, while political writings and dictionaries were published in increasing numbers.[1] In the same way other scholars have shown how the countries of origin of foreign books reviewed in a major French periodical altered during the first half of the century. (There was a marked fall in the proportion of German books and a rise in that of English and Italian ones.)[2] There are considerable technical difficulties to be overcome in studies of this kind. Inventories after death (an obvious source of information about what books were widely owned) often do not list them in any detail if they were unlikely to fetch much when sold. The official sources used by Estivals and Furet do not tell the researcher how large the edition of any particular title was, or sometimes even how many editions there were, and take no account of books published in French in the Dutch Republic, Switzerland, or even the papal enclave of Avignon.[3] Moreover it is, in the nature of things, almost impossible to find out with real certainty what, if anything, the peasant or artisan read, what books the pedlar carried in his pack. Often, again, quantatitive analysis, in this as in other fields of history, merely gives greater precision, at the cost of much effort, to what was already known or at least assumed. It is still possible to write excellent studies of many aspects of French intellectual history without the use of such methods. Louis Trenard, for example, in the first volume of his *Lyon de l'Encyclopédie au préromantisme* (2 vols., Paris, 1958) was able to write a very full study of the intellectual life of the second largest city in France in the years 1770–93, covering such topics as newspapers, book publishing, the theatre, education, and freemasonry, without any use of statistical material or techniques. Perhaps in the long run changes in mental attitudes and assumptions may be best studied by the use of

[1] F. Furet, 'La "librairie" du royaume de France au 18e siècle', in G. Bollême and others (eds.), *Livre et société dans la France du XVIIIe siècle.*

[2] J. Ehrard and J. Roger, 'Deux periodiques français du 18e siècle: "Le Journal des Savants" et "Les Mémoires de Trevoux" ', in *Livre et société . . .'*

[3] See the caveats of D. Ligou, 'Le part du roman dans quelques bibliothèques du XVIIIe siècle', in *Roman et lumières au XVIIIe siècle* (Paris, 1970), pp. 48–63.

linguistic methods, particularly since the computer has now greatly eased the drudgery of this type of analysis. In the mid-1960s, indeed, a linguistic analysis of the titles of all the French books published during the eighteenth century was begun under the auspices of the École practique des hautes études.

It will be clear from the preceding pages that one result of the wonderful growth, during the last three or four generations, of studies of eighteenth-century France has been a certain disintegration of the subject. Precision and completeness have been gained in specialized works, even to some extent in the case of the great regional studies, at the cost of a narrowing of range and scope.[1] General treatments of the subject as a whole, though some excellent ones have appeared,[2] are now, such is the richness of the specialist literature, more difficult to write than ever before. The disintegrating effects of specialization are at least as visible in the writing of the administrative history of the period as in that of its social, economic, or intellectual life. It was much easier sixty years ago for Paul Viollet to attempt, in his *Le Roi et ses ministres pendant les trois derniers siècles de la monarchie* (Paris, 1912) a comprehensive history of the entire central administration than it would be today. The limited range of materials which he used (largely works of political theory and legislative enactments) would now seem utterly inadequate to a professional historian. The present-day tendency, which will certainly persist for the foreseeable future, is towards detailed studies of specific and often quite small administrative groups. In particular the administrative nobility, the *noblesse de la robe*, has attracted a good deal of attention. Jean Egret produced over twenty years ago an important short study of the *parlementaires* and has published a more extensive one of their political significance during the reign of Louis XV.[3] Almost simultaneously Franklin Ford illustrated brilliantly the speed and completeness with which the high *noblesse de la robe* became fused with the older French nobility during the generation after 1715.[4] More recently Fernand Bluche has written a

[1] See the comments of Betty Behrens, ' "Straight History" and "History in Depth" ': the experience of writers on eighteenth-century France', *Historical Journal*, viii (1965), 118.

[2] See above, p. 52, fn. 4.

[3] 'L'Aristocratie parlementaire française à la fin de l'ancien régime', *Revue historique*, ccviii (1952), 1–14; *Louis XV et l'opposition parlementaire* (Paris, 1970).

[4] F.L. Ford, *Robe and Sword: the Regrouping of the French Aristocracy after Louis XIV* (Cambridge, Mass., 1953).

series of works which analyse in detail the social composition of several of the most important French administrative institutions.[1] These assemble a great amount of precise information on such issues as the social status and background of members, their age at entry, and the relative status and sometimes conflicting attitudes of different institutions. These books, which are an approximation for the French administrative system to the type of social analysis which Namier earlier provided for the British House of Commons,[2] expose clearly the extent to which the *noblesse de la robe*, normally referred to in textbooks as a unity, was really split into sub-groups whose outlooks were sometimes in violent conflict.[3] A very similar attempt to place well-defined administrative groups in their social context underlies two other outstanding modern studies. These are Vivian Gruder, *The Royal Provincial Intendants: A Governing Elite in Eighteenth-century France* (Ithaca, 1968), which is based on an analysis of the intendants at three moments during the century—in 1710–12, 1749–51, and 1774–6; and Yves Durand, *Les Fermiers généraux au XVIIIe siècle* (Paris, 1971). The latter in particular, with its heavy emphasis on the importance for this group of family ties and interrelationships, shows strong Namierite tendencies.

The tendencies which have been sketched in the last few pages have been for a generation or more the most dominant and most interesting ones in the historiography of eighteenth-century France. There have, however, been two others which call for at least brief mention. The first of these is a marked growth of interest in the corporate institutions, from the *parlements* and provincial estates at one end of the social spectrum to the trade guilds and *compagnonnages* at the other, which remained important in French life down to the Revolution. This interest is the French aspect of a European trend. The later nineteenth and early twentieth centuries produced a large amount of writing, particularly in Germany, on the corporative society of the later Middle Ages and the early modern period; and by the 1920s and 1930s scholars such as Otto Hintze were beginning

---

[1] *Les Magistrats du parlement de Paris au XVIIIe siècle* (Paris, 1961); *Les Magistrats du grand conseil au XVIIIe siècle, 1690–1791* (Paris, 1966); *Les Magistrats de la cour des monnaies de Paris au XVIIIe siècle, 1715–1790* (Paris, 1966).

[2] See below, pp. 223–6.

[3] For example the Grand Conseil was less politicized, less Jansenist, and more amenable to royal power than its great rival, the *Parlement* of Paris (*Les Magistrats du grand conseil*, pp. 40–1).

to extend comparative studies of this kind to cover the whole of Europe.[1] In 1936 the Comité internationale des sciences historiques, meeting at Bucharest, set up in response to this growing interest a Commission internationale pour l'histoire des assemblées d'états. Most of this activity was concentrated on periods earlier than the eighteenth century; nevertheless it has left distinct traces on writing about France during the last generations of the old regime. In particular F. Olivier-Martin, in his *Histoire du droit français des origines à la révolution* (Paris, 1951) has shown convincingly how much French society and administration during this period remained dominated by corporate institutions inherited from previous ages.[2] Working on a more restricted canvas, Émile Coornaert has shown, in his *Les Corporations en France avant 1789* (Paris, 1941), the continuing power of guild traditions and attitudes even in trades where guilds themselves did not exist. He emphasized what was already fairly well established—that corporate bodies of this kind were at the height of their importance in France during the two generations or more before 1735–40, the years which saw the government for the first time begin to discourage the formation of new trade guilds. In 1729–60 the *Bureau du commerce* rejected 238 requests for the establishment of new bodies of this kind; and even in 1789 there were almost as many favourable as unfavourable judgements of the guild system to be found in the *cahiers* of the Third Estate.[3] The same author has shown, in a book of great interest, how the *compagnonnages*, the traditional associations of workers now deeply rooted in a number of skilled trades, grew considerably in numbers in eighteenth-century France and became increasingly tightly organized as corporate institutions (for example in the increasing use by many of them of seals with which their acts were authenticated).[4] He also shows how completely traditional was the outlook of these associations, which traced their origins in most cases to the building of Solomon's temple and admitted new members only to the accompaniment of complex pseudo-religious rituals.[5]

---

[1] E. Lousse, *La Société d'ancien régime: organisation et représentation corporatives* (Louvain–Bruges, 1943), pp. 35 ff., 41.

[2] Particularly in Book II, chap. ii, 'La Nation organisée'. See also the collective work, *Les Étapes de la législation corporative en France* (Paris, 1944).

[3] 2nd edn. (Paris, 1968), pp. 146, 150, 167, 173.

[4] E. Coornaert, *Les Compagnonnages en France, du moyen age à nos jours* (Paris, 1966), p. 49.

[5] pp. 22–4, 153–68.

Such writing is a valuable corrective to the deceptively rational and materialist picture painted by the movement of quantitative analysis which traces from Labrousse. This interest in the corporate aspects of old-regime society has sometimes been inspired, like the materialist and quantitative approach, by political presuppositions (the Vichy government established an Institut d'études corporatives, though this produced little of value). Nevertheless this line of approach has added a valuable strand to the increasingly complex fabric of the historiography of eighteenth-century France.

There has also been, from the 1950s onwards, an effort to place French society and its problems during the eighteenth century more integrally than before in their European context, to stress the extent to which the tensions which erupted in France at the end of the 1780s can be paralleled over much of western, and to a lesser extent central and eastern, Europe. The French Revolution, it is argued, was merely the greatest and most violent manifestation of an 'Atlantic Revolution'. This, directed above all against the structure of corporate privilege by which existing societies and governments were more and more dominated, was to be seen, it is claimed, in many parts of Europe and even across the Atlantic in the movement for independence in the British American colonies.

The idea remains a highly controversial one. Its fullest and best-known statement, by the American R. R. Palmer in his *The Age of the Democratic Revolution* (2 vols., Princeton, 1959–64) has not won the general assent of historians.[1] Palmer agrees that the revolutions or attempted revolutions of the later eighteenth century—in North America from 1775 onwards; in Geneva in 1782; in England (in the form of demands for parliamentary and 'economical' reform) in the early 1780s; in the Dutch Republic and the Austrian Netherlands in the later 1780s; in France from 1789 and in Poland from the meeting of the 'three-year diet' in 1788—had no single centre and differed widely between themselves. There is no question here of the revival of myths of a ramifying international conspiracy for the subversion of the social order. Nevertheless it can be argued strongly that the picture Palmer draws is too exclusively dominated by politics and particularly by political ideas. The fact that nearly all the movements he describes sprang from similar political and intellectual

---

[1] There are convenient summaries of Palmer's main conclusions in his 'The World Revolution of the West, 1763–1801', *Political Science Quarterly*, lxix (1954), 1–14; and in his 'Reflections on the French Revolution', ibid. lxvii (1952), 64–80.

roots is elaborated interestingly and at length. The fact that they developed in radically different economic and social environments is less emphasized. Yet his argument has a long and respectable ancestry; it is interesting that as early as 1789 at least one French pamphleteer was already grouping together the American, Genevan, and French revolutions since all, he alleged, were inspired by the idea that 'the law is the act or expression of the General Will' and thus traceable ultimately to Rousseau.[1] Moreover the 'Atlantic Revolution' thesis has found followers in France. There Jacques Godechot had already adopted it before it had been fully developed in English and has over the last two decades popularized it in several widely read books.[2] Even though this approach to the later eighteenth century may in the long run be found, like so many historical ideas, too simple and too schematic, it has none the less played a constructive role in encouraging comparative studies of European society and government in the later eighteenth century. Serious work of this type is now more common and more fruitful than ever before.[3]

Even so summary a view as this of the historiography of eighteenth-century France must end on a note of pride and optimism. Between the writing of the early nineteenth century and that of today, the gulf is enormous. The first was for the most part narrowly political and indifferent to events outside Paris, often envenomed by party prejudice; the second, with all its faults, is sophisticated, many-sided, and technically polished to an extent hardly imaginable even a few decades ago. It is still possible for mediocre accounts of the subject, uninfluenced by the great advances made since the First World War, to appear[4] and even for viciously biased and xenophobic accounts of the origins of the Revolution[5] to find publishers. But the advances,

[1] A.-N. Isnard, *Observations sur le principe qui a produit les révolutions de France, de Genève et d'Amérique dans le dix-huitième siècle* (Evreux, 1789). The most complete account of the intellectual influences which flowed from the newly independent America to France in the years before 1789 is still B. Fay, *L'Esprit révolutionnaire en France et aux États-Unis à la fin du XVIIIᵉ siècle* (Paris, 1925), bu this lacks the schematic qualities of Palmer's book.

[2] *La Grande Nation: l'expansion révolutionnaire de la France dans le monde de 1789 à 1799* (2 vols., Paris, 1956), i. chap. i; *Les Révolutions, 1770-1799* (Paris, 1963).

[3] A good example is the proceedings of the Colloque sur l'abolition du régime féodal dans le monde occidental, covering the whole of Europe in the later eighteenth century, printed in *Annales historiques de la révolution français*, N.S., xli (1969), 145-371.

[4] e.g. P. Gaxotte, *Histoire des Français* (Paris, 1951), vol. ii.

[5] e.g. B. Fay, *Louis XVI, or, The End of a World* (London, 1968; French edition, Paris, 1966).

in sympathy and width of view as well as in mere factual knowledge, have been enormous. We not only know more about the last generations of the old regime in France than Michelet, Taine, or even de Tocqueville: we see them in a different way, broader, more complete, even more humane. The lesson is clear. We are not greater historians or deeper scholars than the predecessors on whose work we have built. But we have, little by little, by painful accretions, built up an understanding of the past which is qualitatively as well as quantitatively superior to any they possessed.

CHAPTER II

THE ENLIGHTENMENT

THE EIGHTEENTH-CENTURY Enlightenment is a subject full of pitfalls and subtleties, above all because it is so vast in scope and so amorphous. It embraces a wide range of differing, sometimes radically conflicting, tendencies. Were its dominant characteristics rationalism and optimism? Perhaps; but it also included active and powerful elements of doubt, of pessimism, of romanticism of different sorts and colours. To speak of the Spirit of the Age in any historical period, especially in one of change and development, must always involve a gross and more or less arbitrary approximation to the truth; and this is at least as much true of the eighteenth century as of any epoch in modern history. There is an obvious danger that any writer on the subject may seize, at the dictates of his own personality and prejudices, on one facet of the intellectual life of the period and proclaim it as the most significant or even the only one of significance. There is a second and even more powerful temptation, more marked perhaps with regard to the Enlightenment than to the subject matter of any other chapter in this book; that of interpreting the past under the influence of present events, especially when these are as spectacular and frightening as the upheavals of the French Revolution and the outbreaks of 1848 and 1871. Finally there is the difficulty, inherent in the writing of all intellectual and cultural history, of distinguishing between ideas and the words which express, but also sometimes blur and confuse, them. Every age has its weapon-words, terms which are worn threadbare in controversy and which are often in the first place only very approximate and ambiguous expressions of the idea they attempt to convey. 'Nature' and 'Reason' are obvious examples of such terms in the eighteenth century; irreconcilably different thinkers and writers could claim, quite sincerely, to base themselves on these cloudy concepts.[1]

The subject then is not an easy or straightforward one. Until within

---

[1] There is a good discussion of these difficulties in G. Boas, 'In Search of the Age of Reason', in *Aspects of the Eighteenth Century*, ed. E. R. Wasserman (Baltimore–London, 1965), pp. 1–19.

the last generation many of its subtleties have not been clearly seen and many of its pitfalls have not been avoided by historians. Writing on the Enlightenment during the eighteenth, nineteenth, and even much of the twentieth centuries has been deeply marked by partisanship, by views narrowed by political, religious, and national prejudice. The rise in intellectual standards, the growth in scope, in impartiality, in originality, and in sheer erudition, which has marked the study of the eighteenth century during the last thirty or forty years has been perhaps more marked in this than in any other field.

TRADITIONAL VIEWS: AN AGE OF REASON AND FREEDOM
OR ONE OF MERELY DESTRUCTIVE CRITICISM?

If the summit of the Enlightenment be taken as the years around or just after 1750, the period which saw the publication of the most important works of Montesquieu and Condillac, of perhaps the boldest work of Diderot (his *Lettre sur les Aveugles* of 1749), of the first volumes of the *Encyclopédie* and the earliest significant writings of Rousseau, then for not far short of two hundred years to come assessments of it tended to follow one or other of two radically divergent paths. On the one hand, it was argued that it had been a period of liberation of the human mind, an age marked by the breaking of intellectual and therefore, in the long run, of political, social, and economic fetters, by a growth of rationality in thought and therefore in action, by genuine progress and justified belief in progress. On the other, the Enlightenment was denounced, and many of its leading figures often violently attacked, as expressing a rationalism which was arrogant, superficial, and totally unrealistic. It was scorned by this second current of thought as an irresponsible undermining by conceited and self-willed men of the essential bases of society and civilized life. These deep-rooted and traditional attitudes are far from having disappeared today. The passage of time and a great outpouring of detailed research during the last generation has done something to reduce the acerbity with which they are voiced and to introduce an increasing range of grey tones into a previously black and white picture. But the voices of what may, to use inevitable shorthand terms, be called the idealistic optimist radical and the realist pessimistic conservative are still almost as audible as they have ever been.

The first of these attitudes was that of most of the more radical writers of the Enlightenment, above all in France. These men in general believed that all human progress depended upon the development and spread of habits of rational thought. For this mathematics and the mathematics-based physical sciences were the obvious model; and in face of these, traditional modes of thinking and a reliance on tradition and the past must be discarded.[1] Some at least of these thinkers believed in an optimistic and even mechanistic way in the reality of human progress,[2] while so great a thinker as Kant saw the essence of the Enlightenment in highly optimistic terms as the achievement by man at last of a maturity defined as the ability to use his mind freely and autonomously, without reliance on outside guidance.

Moreover in the nineteenth century the question of the character of the Enlightenment became bound up, apparently indissolubly, with that of the rightness and justification of the Revolution of 1789. It was very widely taken for granted that the Enlightenment had been a major cause of the Revolution; at the very least it had done much to prepare the way for it in men's minds. The two then seemed to stand or fall together. A liberal, a radical, a secularist, later in some cases even a socialist, who approved of the Revolution as the greatest step forward ever taken by humanity, was therefore bound to believe in the truly rational, progressive, and liberating character of the movement of ideas which had preceded it. Thus in the 1820s the philosopher Victor Cousin saw the Enlightenment as carrying out above all a necessary mission of destruction, as disposing at last of the inheritance of the Middle Ages which had for centuries weighed so heavily on the human mind. This it had done by its hostility to all traditional authority in the realm of thought, and particularly by its rejection of hypotheses in favour of rigorous

---

[1] As in Diderot's assertion that 'Physical and mathematical proof should take precedence of moral proof, just as the latter should take precedence of historical proof' (*Oeuvres*, Paris, 1875–7), ii. 81. In the same way Dupont de Nemours defended the intellectual status of economics by claiming that its propositions were 'as severe and incontestable as those of geometry or algebra' (A. Mathiez, 'Les Doctrines politiques des physiocrates', *Annales historiques de la révolution française*, xiii (1936), 193). Similar illustrations could be multiplied indefinitely. On the influence of science and technology on the French vocabulary in the eighteenth century see F. Gohin, 'Le Mouvement des idées et les vocabulaires techniques au XVIIIᵉ siècle', *Revue historique*, clxviii (1939), 307–20.

[2] As in Condorcet's belief that the achievement of American independence, by doubling the number of those who devoted themselves to progress, must also double its speed ('De l'influence de la révolution de l'Amérique sur l'Europe', in *Collection des économistes*, xiv (Paris, 1847), 558).

analysis. 'The political mission of the eighteenth century', he wrote, 'was to break with authority; its more special mission, where method was concerned, was to break with the hypothesis'. He agreed that its achievement had been negative. This was inevitable, for 'a century, a single century, can scarcely be charged with two missions simultaneously. It destroyed, it created nothing: but it could do no more'. Its legacy was one only of great abstractions; but these 'are immortal truths in which the future is contained'.[1] On a more popular and less purely philosophical level rather similar ideas were put forward by Pierre Lanfrey in the 1850s. Above all the tone is similar; and it is this tone—abstract, high-minded, and dogmatic, valuing intellect rather than imagination or sensibility, with a distinctly anticlerical and sometimes nationalistic note—which is audible in most of the favourable comment on the Enlightenment throughout the nineteenth century. 'Heir of the sufferings but also of the experience of its predecessors,' Lanfrey wrote of the eighteenth century, 'it set itself the mission of carrying the light of analysis, of reflection and of reason into the grave problems of which the solution had hitherto been entrusted only to the edge of the sword or the vain theories of the imagination.' It had banished metaphysics, with all its confusing and destructive effects, from philosophy. It had thus brought the latter down to earth and made it of practical use. 'This transformation, this renaissance, this sovereignty of reason, this marriage of philosophy with the realities of life, is the greatest fact of modern history.' As a result of the work of the Enlightenment, by the end of the eighteenth century the church and all it stood for had been defeated almost everywhere in Europe; so that 'infallible orthodoxy held its ground only in a few remote cantons of Spain and Italy', while 'Rome stood motionless, stupefied and amazed by the defection of the peoples, without making any attempt to rally them around her'.[2]

Such statements of the liberal and radical view of the Enlightenment were most frequent and most forthright in France; but they were by no means confined to one country. It had been a genuinely international movement (to a degree, indeed, underestimated until our own day); its nineteenth-century advocates therefore could be

---

[1] *Cours de philosophie* (Brussels, 1840), i. 35, 64, 100–1, 108. This book prints a course of lectures which Cousin gave in 1829.

[2] *L'Église et les philosophes au dix-huitième siècle*, (Paris, 1879: the book first appeared in 1855), pp. 134–5, 362–3. See also Lanfrey's *Essai sur la révolution française* (Paris, 1858), pp. 41 ff.

found throughout the civilized world. A good English example is the Gladstonian liberal John Morley who in his *Diderot and the Encyclopaedists* (London, 1878) took a viewpoint more typically French than English. In particular he exemplifies, to a degree unusual in the English-speaking world, the anticlericalism which was often one of the most obtrusive ingredients of French liberal writing on the Enlightenment. The French Church in the eighteenth century he identified simply and unhesitatingly with 'superstition, ignorance, abusive privilege, and cruelty'. Against it the Enlightenment and, above all, the contributors to the *Encyclopédie* had mounted an offensive inspired both by the intellectual stimulus stemming from scientific discovery and by hopes of a more just and rational society. In doing this they had pointed out the only possible road of progress; and after an unfortunate reaction against the *philosophes* inspired by the Revolution of 1789 and some of its results mankind was once again treading this road. 'Materialistic solutions in the science of man, humanitarian ends in legislation, naturalism in art, active faith in the improvableness of institutions—all these are once more the marks of speculation and the guiding ideas of practical energy.'[1]

Throughout the nineteenth century, then, this liberal attitude was entrenching itself and becoming an orthodoxy to its adherents. Attacks on it by conservatives and clericals were expected, and rejected when they were made. Questioning of it by scholars, attempts to show that the facts did not bear it out, were sometimes less easy to deal with but were also rejected. Thus when Edmé Champion, in his *La France d'après les cahiers de 1789* (Paris, 1897) showed that the *cahiers* contained few echoes of the ideas of the Enlightenment and argued that the Revolution had been the result merely of concrete grievances and above all of the food shortage of the spring and summer of 1789, this idea was at once rebutted. There had, it was argued with a good deal of force, been less physical suffering in 1789 than in the famine of 1709 or even the difficult year of 1753; yet neither of these had seen a revolution. What was new in 1789 was the vision and the intellectual stimulus provided by the *philosophes*.[2] In the twentieth century the idea of the Enlightenment as the sub-

[1] i. 2, 8.

[2] See e.g. M. Roustan, *Les Philosophes et la société française au XVIIIᵉ siècle* (Paris, 1911), pp. 13–15. The idea that the *philosophes* had had little to do with the making of the Revolution had been earlier put forward by Félix Rocquain in his *L'Esprit révolutionnaire avant la révolution, 1715–1789* (Paris, 1878).

stitution of the liberating principle of reason for the confining and stultifying ones of tradition and authority has continued to find eminent adherents. In an intelligent general survey written in the 1920s Henri Sée put forward this idea essentially in the form which it had taken for generations.[1] A decade later, in a more wide-ranging book, the American Preserved Smith claimed, in rhetoric which Lanfrey would not have disowned, that 'The cloud of superstition which had darkened the Middle Ages . . . rolled away when the sun of reason rose and shone with all its strength.'[2] But the most sustained and systematic defence of the Enlightenment against its critics by any twentieth-century writer has been that offered by Peter Gay. For two decades, in an impressive series of books and articles, he has argued that its essence was freedom in all its forms, 'freedom, in a word, of moral man to make his own way in the world', and that it was neither foolishly optimistic nor irresponsible, as its opponents so often asserted.[3] It criticized in order to clear the way for positive action, and thus gave criticism a truly constructive role. The *philosophes* were neither silly utopians despising the past nor superficial believers in easy progress; on the contrary they 'were often pessimists, usually empiricists, generally hard-headed political men, with sensible programs, limited expectations, and a firm grasp of history'.[4]

Against this current of praise and approval ran from the beginning another, equally strong and deep, of hostility and often of bitter denigration. In the very first years of the nineteenth century J. L. Soulavie, in his *Histoire de la décadence de la monarchie française* (Paris, 1803) attacked violently the disruptive and demoralizing effect of the work of the *philosophes*, and their conceit and intolerance. He also levied against the Enlightenment an accusation to be frequently repeated for a century or more; the charge that it had undermined healthy French national feeling and had

[1] *L'Évolution de la pensée politique en France au XVIIIᵉ siècle* (Paris, 1925), pp. 342–3, 351, 383.

[2] *A History of Modern Culture*, ii. *The Enlightenment, 1687–1776* (London, 1934), p. 361.

[3] *The Enlightenment: An Interpretation. The Rise of Modern Paganism* (London 1967), p. 3 and Preface, *passim*.

[4] *The Party of Humanity: Studies in the French Enlightenment* (London, 1964), p. 262. For a short but forcible earlier statement of Gay's ideas on this point see his 'The Enlightenment in the History of Political Theory', *Political Science Quarterly*, lxix (1954), 374–89. A good modern expression of the more traditional view of the Enlightenment as lacking in true historical sense and any concept of organic growth can be found in R. N. Stromberg, 'History in the Eighteenth Century', *Journal of the History of Ideas*, xii (1951), 295–304.

replaced it only by a flabby cosmopolitanism. It had failed to support Louis XV and Louis XVI. It had helped to create an 'English faction' which had been powerful enough to secure the dismissal of Turgot in 1776 and to transform the constructive national revolution of 1788 into the destructive and foreign-inspired one of 1789.[1] No more than the liberals and radicals did Soulavie doubt that the Enlightenment had paved the way for the Revolution; but since he disapproved of the result he must also disapprove of the force which had produced it. For fifty years the *philosophes* had 'taught the theory of what must be done to decompose the supreme power . . . with such effect that there is no terrible event offered by the revolution the precepts of which you will not find word for word in their writings'.[2] This is a long-winded, poorly-organized, and unintelligent book. In it Soulavie's search for a scapegoat, for preference a foreign one, on whom responsibility for the outbreak of the Revolution can be placed is very much in evidence. The book is also deeply marked by the anglophobia so strong in France when he wrote, and by the popularity of vague and emotionally expressed conspiracy theories as an explanation of the catastrophes which had overtaken the country. Not all conservative criticism of the Enlightenment, however, was so unbalanced and unrealistic. Under the Restoration Charles Lacretelle, in his *Histoire de France pendant le dix-huitième siècle* (Paris, 1819) was able to take a relatively fair and moderate view. He had no doubt of the superficiality of the Enlightenment's belief in progress. 'Men coldly intoxicated made a thousand prophecies of the happiness of humanity; honesty, honour, public spirit, the love of humanity, seemed such simple things that men tried to draw up rules for them as for an arithmetical calculation'. He also saw that the contempt for pedantry and mere erudition felt by the *philosophes* had made them unfairly suspicious of deep study of any subject, while an emphasis on wit and good conversation, and the whole tone of educated society, had tended to weaken and undermine religious belief. 'Intelligence attempted to find cures for each of the ills which afflict men; and nevertheless religion, which assuages these ills better than anything, was being destroyed.' Yet he realized that many of the *philosophes* had been men of real modesty and wisdom, that the eighteenth century had been a period of real

---

[1] iii. 108–9, 122. There are further vague denunciations of the 'faction presbytérienne de la Grande-Bretagne' at iii. 409, 425.

[2] iii. 397.

intellectual growth, above all in the physical sciences. Of the more radical and speculative of the *philosophes*—Rousseau, Mably, Raynal, Diderot—he disapproved; yet there were others, notably and predictably Montesquieu, of whom he had a high opinion.[1] A few years later Guizot was equally moderate in his criticisms. To him the real weakness of eighteenth-century intellectual life in France had been its purely speculative character and its complete separation from action and the demands of the real world. This 'gave to the ideas of the time a singular character of ambitiousness and inexperience; for never was philosophy more eager to rule the world, or at the same time less conversant with it'.[2]

Neither of these writers, however, was typical of the period in which he wrote. The Restoration was an age of violent controversy in French historical writing, one in which the liberal and the monarchist–clerical views of France's past clashed sharply. The Revolution was still too close in time, memories of it too vivid, for its opponents to take a balanced view of the great intellectual movement which was so widely regarded as its most important cause. A similar colouring of assessments of the Enlightenment by the events of the recent past can be seen in the years after 1870. The catastrophe of that year seemed to throw into sharp relief weaknesses of French intellectual life and the resulting defects of society in general. These weaknesses, it could be argued, were largely the legacy of the Enlightenment. Many of the judgements passed on it during the early years of the Third Republic were therefore markedly hostile. Taine, with his active sense of the faults and weaknesses of the ordinary man, his deep distrust of 'la raison raisonnante', is the outstanding illustration of this.[3] Almost simultaneously with the publication of the first volume of his work, moreover, another author of real ability was attacking the intellectual life of the eighteenth century, from a different standpoint, because of its weakening and corrupting emphasis on happiness and because of its hatred of Christianity, 'this religion which had above all in its eyes the fault of being sad'.[4] Nor did Taine in his attack on the *philosophes* lack disciples in the later decades of the century. André Lichtenberger, in his *Le Socialisme*

---

[1] iii. 91–2, 98, 129, iv. 122.

[2] H. F. Guizot, *General History of Civilization in Europe, from the Fall of the Roman Empire till the French Revolution* (Edinburgh, 1848), p. 239.

[3] See above pp. 29–33.

[4] A. A. Cournot, *Considérations sur la marche des idées et des événements dans les temps modernes* (Paris, 1872), ii. 59–60, 67–8.

au XVIIIe siècle (Paris, 1895) discussed the Enlightenment in very much the same terms (he frequently refers to Taine in his footnotes). Like his great predecessor he criticized it as dominated by a belief in pure and abstract reason of an essentially mathematical type and by an *esprit classique* which was indifferent to or ignorant of the real world and which 'loved maxims which were simple and general in scope'.[1]

All these writers, whether partisans or opponents of the Enlightenment, wrote from a philosophical or political standpoint. By the last years of the nineteenth century and in the early years of the twentieth the subject was being seriously discussed for the first time from a different point of view, that of the historian of literature; and this discussion was in general markedly hostile in tone. Hitherto the literature of eighteenth-century France had aroused little scholarly interest compared with that of the age of Louis XIV or even that of the sixteenth century. Now a number of eminent critics and literary historians—notably Émile Faguet and Ferdinand Brunetière—were beginning serious study of this aspect of the period and returning unfavourable verdicts on it. To Faguet the eighteenth century appeared 'remarkably pale between the age which precedes and that which follows it'. It had witnessed a distinct decline of the moral sense in France, 'the brusque extinction of the Christian idea' and a marked weakening of any feeling of patriotism. Since it lacked a tradition, or even any desire for one, it had been 'a childish, or if you like adolescent, century'. The *philosophes* had been too conceited and too much involved in the world of affairs (here of course he contradicts diametrically the more common hostile view of them as impractical and remote from the real world) to be really serious in an intellectual sense. They were mere polemicists rather than philosophers; they 'never see further at any given moment than their immediate idea to prove and their adversary to confound'. The literature of the period had been neither truly innovating nor truly traditional, and had been weakened above all by its lack of any national character. In any case the history and physical sciences which the century did so much to develop had themselves shown the falsity of its ideas. History teaches that living traditions are of fundamental importance to a people; while science has exploded the idea of equality, since the workings of natural selection show that race and aristocracy are proved facts. Political science, also largely a creation

[1] pp. 3–5, 28.

of the eighteenth century, shows that real progress is to be made by observation and measurement, not by the propounding of theories or the recital of syllogisms.[1] In any case this intellectually mediocre Enlightenment had had no influence whatever as a cause of the Revolution, which had been the product of poverty and hunger: the 'principles of 1789' were a myth.[2] This line of argument, which regarded the Enlightenment as not so much wicked as hopelessly second-rate, doomed to failure because in conflict with the real nature of society and human life, was not entirely new. But it had seldom been put so mercilessly.

Brunetière shared this low view of the literature of eighteenth-century France and saw the period, like Faguet, as one of moral decline. Since really fundamental moral questions were impossible to answer with the certitude which was now attainable in the physical sciences, they had tended to be neglected, almost to evaporate. Above all, however, the great error of the eighteenth century had been its overvaluation of sensibility and the idea of nature, which led inevitably to the triumph of individualism. If everything is natural, the distinction between right and wrong disappears; and in any case art and the higher aspects of life are not natural at all.[3] The *philosophes* (of whom Brunetière regarded the contributors to the *Encyclopédie* as the supreme representatives) had no precise ideas on many fundamental political questions, such as the nature of sovereignty, the division of powers, or the extent of the rights and duties of the state. Indeed the dominant characteristic of their thinking was a marked lack of clarity; in this they compared very unfavourably with the great writers of the seventeenth and nineteenth centuries.[4] Like Faguet, Brunetière found it easy to denigrate the men of the Enlightenment by comparing them unfavourably with their predecessors and successors. Less than any other scholarly critic of the Enlightenment did he disguise his personal beliefs and the extent to which they dominated his attitude to the subject. He opposed any idea of political equality as contrary to justice, an impediment to liberty, and in any case impossible to maintain for

---

[1] *Dix-huitième siècle: études littéraires* (Paris, 1890), Avant-Propos, *passim*, especially pp. v–vi, xii–xiii, xx, xxix–xxx.

[2] 'La France en 1789', in *Questions politiques* (Paris, 1899), especially pp. 3, 5, 8, 19. This article is a reprinted review of Champion's book.

[3] *Histoire de la littérature française classique (1515–1830):* vol. iii. *Le Dix-huitième siècle* (Paris, 1912), pp. 322–9, 332, 364–7.

[4] pp. 392, 411–12.

long. To him social equality was a chimera, but one capable of ruining the family and the 'superior parts' of a nation or of humanity in general.[1]

As with Brunetière and such predecessors as Taine, so with many of his successors. A critical attitude to the Enlightenment, sometimes bitterly hostile, sometimes merely condescending, has almost always been the product of cynicism towards the ideals which it proposed and pessimism as to the reality of the hopes which it cherished. In our own day, however, these attitudes have been sharpened by a fear unequalled since the early decades of the nineteenth century of the long-term implications of its more radical aspects. The chaos from which Europe has suffered during the last two generations, the terrifying destructive potentialities which the period has revealed in what were thought of as civilized societies, have made at least certain aspects of the Enlightenment much more than the object of merely academic discussion. Taine indeed began his *Origines de la France contemporaine* in an atmosphere of fear and pessimism engendered by national defeat and the apparently imminent danger of domestic revolution; but the writers of the later nineteenth century, whatever their attitude, worked in an environment of social stability and economic security. Since 1914, and above all since 1917, this has no longer been so. The change has meant that serious academic criticism of the Enlightenment has tended increasingly to centre around the contribution which some aspects of that movement unquestionably made to many forms of twentieth-century totalitarianism in general and to Bolshevik idealism in particular. More and more the charge has been not that the Enlightenment was ineffective but that it was, at least in the long run, all too successful; not that it was intellectually second-rate but that it created a destructive new tradition in European political thought and feeling.

The most famous and influential attack on the potentially totalitarian aspects of the Enlightenment is that made by J. L. Talmon in his *The Origins of Totalitarian Democracy* (London, 1952). His central argument is that the eighteenth century, which gave rise to the liberal form of democracy, also generated a totally different totalitarian type. This thought of politics not in terms of trial and error, of a pragmatic adjustment of forms to changing necessities, of political and social systems as merely human productions and therefore

[1] p. 517.

always less than perfect, but rather in terms of 'a preordained, harmonious and perfect scheme of things, to which men are irresistibly driven, and at which they are bound to arrive'.[1] During the second half of the eighteenth century the dream of constructing a planned and uniform society, one which would be natural and rational, and therefore, as the only really valid social system, unchanging, became increasingly widespread and attractive. Such a vision ruled out totally any diversity of view and interests of the kind assumed by liberal democracy. Instead it emphasized the necessity of unity and unanimity, which were to be produced by a willing conformity, fostered by the legislator, to the uniquely valid system which the totalitarian vision postulated. The Enlightenment, Talmon insists, was deeply hostile to any idea of freedom as dependent on the conflict of ideas and interests and compromise between them; and the consolidation of this attitude he attributes particularly, and with justice, to Rousseau.[2] In psychological terms, the most influential idea of this entire radical-enlightened current of thought was that of virtue, which continually and intensely preoccupied many of the writers of the Enlightenment. This virtue, defined in terms which were rigid, puritanical, sometimes positively ascetic, and often highly conservative, bred a general fear and contempt for trade, for large and ostentatious accumulations of wealth, for big cities and sophisticated urban civilization in general. Having outlined these essentially totalitarian ideas and attitudes in the first part of the book, Talmon goes on to discuss in some detail the two greatest efforts of the later eighteenth century to apply them in practice, the Jacobin dictatorship of 1793–4 and the pathetically unsuccessful *Conspiration pour l'Égalité* led by Gracchus Babeuf in 1796. In both cases he emphasizes the harshness and rigidity, the cutting edge of intolerance and fanaticism, given them by the intellectual soil in which they had been nourished. The book is not, and does not attempt to be, a systematic account of the Enlightenment as a whole. It is concerned merely with political and social thought of a certain kind, not with the intellectual life of the eighteenth century in general. Even within its own limits it is open to criticism. It says, for example, little about the origins of the attitudes it describes with such force. They seem to emerge, in Rousseau, Helvétius, Holbach, Morelly, and Mably, with curious suddenness and little intellectual

[1] pp. 1–2.
[2] p. 44, and chap. iii, *passim*.

ancestry. Nevertheless it is a seminal book, and one which could not possibly have been written before 1917 or perhaps 1945. The totalitarian elements in the Enlightenment could not be seen for what they were before the emergence of totalitarianism as an active political and social phenomenon.

No other writer has attacked the radicalism of the Enlightenment with such pungency as Talmon. Nevertheless the conservative assault on it has continued in recent years from a number of different directions. Thus L. I. Bredvold has concentrated his critical artillery above all on its rejection of the idea of Natural Law and its exaltation of that of Natural Rights, independent and self-sustaining. The villain here, however, is not Rousseau but Locke, for 'by this severance of Rights from Law he provided the central doctrine of revolutionary thought for the eighteenth century and down to our own time'.[1] Any weakening of the idea of a Law of Nature, Bredvold argues, must be disastrous, since it 'leads to extreme individualism, to perpetual protest against all tradition, to the rejection of anything that is inhibiting or contrary to our natural impulses. It sets up a profound opposition between Nature and civilization'.[2] The one eighteenth-century thinker whom he speaks of with real enthusiasm is, almost inevitably, Burke, since he rejected the grossly inferior substitutes for the Law of Nature which the Enlightenment had elaborated and reaffirmed the ancient doctrine in all its potency.

But the most large-scale and detailed criticism of many aspects of the Enlightenment is contained in two works by L. G. Crocker which together constitute an imposing study of French ethical thought in the eighteenth century. These are *An Age of Crisis: Man and World in Eighteenth-century French Thought* (Baltimore, 1959) and *Nature and Culture: Ethical Thought in the French Enlightenment* (Baltimore, 1963). The first in particular is a powerful statement of a pessimistic anti-Enlightenment point of view. The essence of the argument is that the *philosophes*, in so far as they deprived ethics of any supernatural sanction and justified them merely in terms of human life, were attempting the impossible. All modern history shows that attempts to construct a rationalistic and moderate ethical system are doomed to failure. Nevertheless the thinkers of the eighteenth century were in some ways only too effective. 'By affirming what they perceived to be man's true place in the universe, they loosed the

[1] *The Brave New World of the Enlightenment* (Ann Arbor, 1969), p. 24.
[2] p. 82.

metaphysical moorings and set him adrift.' Crocker is appalled by such a 'loss of the metaphysical fundament of values and the consequent ethical confusion and uncertainty'. In this he sees 'the opening chapter of the moral crisis of the modern world, which has come to such a critical pass in the twentieth century'; and he argues forcibly that much eighteenth-century thinking tended to lay 'the foundations of moral nihilism.'[1] As in so much writing of the last generation the Enlightenment is here condemned not merely in terms of its intrinsic intellectual quality but above all for what is alleged to be its legacy to the twentieth century; the tendency, always to some extent inescapable, to judge the past in terms of how it is believed to have influenced the present, is very obvious. In particular the *philosophes* are condemned for having overestimated the ease with which men could be brought to subordinate immediate self-interest to the good of society in general (Condorcet is quoted as an outstanding example of this). This led them to underestimate greatly the degree of conditioning and the intensity of control from above which would be needed to produce such a subordination; and this fatal error is illustrated by the increasingly authoritarian character of the French and Bolshevik Revolutions.[2] The case is powerfully argued but certainly one-sided. The current of practical liberal reform (exemplified by Turgot and even Condorcet, for example) which was a considerable element in the Enlightenment is largely ignored, while Professor Crocker is able to see the roots of moral nihilism even in such unlikely places as the works of Pascal (because of his stressing of the frailty and depravity of man) and Montesquieu. In the second book the verdict is somewhat less uncompromising. Nevertheless, after a long discussion of the struggles of the *philosophes* with their clerical and conservative opponents, and with each other, over such moral issues as the role of the passions in an ethical system, the problem of evil, optimism, and the moral potentialities of education, the author is driven once more to the conclusion that the eighteenth century failed to evolve a workable system of merely naturalistic ethics. Much worse, by making the attempt it raised the spectre of moral nihilism. 'Perhaps the greatest lesson of their enterprise,' he concludes of the *philosophes*, 'is that the basic problems cannot be solved in terms of political reforms or enlightenment. The trouble is in us, in the human condition.'[3] This is a verdict which the clerical

---

[1] *An Age of Crisis*, pp. 471, Preface xi, 451–2.
[2] pp. 455–7.   [3] *Nature and Culture*, p. 512.

critics of the *Encyclopédie* or of Voltaire could have accepted without changing a syllable. After more than two centuries the clash of ideas and attitudes persists, at the deepest level virtually unaltered, because it is rooted ultimately in the existence of different sorts of human personality.

VOLTAIRE AND ROUSSEAU:
THE HISTORIOGRAPHY OF TWO THINKERS

This clash, between the vision of the Enlightenment as liberation and the denunciation of it as casting man adrift without the compass which revealed religion or historic tradition in different ways provide, also underlies all discussion of individual thinkers. Of this an inordinately large part has, until well into the twentieth century, been devoted to one man—Voltaire. That a thinker so essentially second-rate should have attracted for generations more attention than Rousseau, the greatest single creator of the modern sensibility, or Montesquieu, arguably the greatest figure since Aristotle in the history of political ideas, or even Diderot, to whom Voltaire was far inferior in intellectual daring and personal attractiveness, is a remarkable fact. Nothing attests better the importance, so far as the ordinary man is concerned, of form and style in the presentation of any system of ideas. Stripped of the wit with which it was expressed, the mediocrity of much of Voltaire's thinking would have been clearer and the impact of his work more muffled. Nevertheless the intensity of the interest he attracted is beyond dispute; around him far more than any other individual assessments of the Enlightenment were centred. A bibliography published in 1929 lists almost 1,500 books and articles published on him during the century 1825–1925 (and this excludes publications in the Slavonic languages). Another which appeared in 1968 includes over 2,000 titles[1] which appeared in the forty years 1925–65. Moreover, of these about two-fifths were produced in the English-speaking countries, a very much higher proportion than in the nineteenth century and an indication of the increasingly wide geographical spread of interest in Voltaire.

His pre-eminence over the other great figures of the Enlightenment, at least until the First World War, in the amount of attention which

[1] Mary-Margaret H. Barr, *A Century of Voltaire Study: A Bibliography of Writings on Voltaire, 1825–1925* (New York, 1929); D. Roche, 'Voltaire, aujourd'hui', *Revue historique*, No. 500 (Oct.–Dec. 1971), 341.

he has attracted from historians can perhaps be illustrated in a different way. The nineteenth century saw the publication of two great editions of his *Œuvres*, those edited by Beuchot (1829–40) and Moland (1877–85), as well as of many editions of individual works, while 107 volumes of his letters appeared under the editorship of Theodore Besterman in 1953–66. Rousseau offers a significant contrast. Though several substantial editions of his works were published before the end of the eighteenth century he attracted very little attention of this kind during that which followed, apart from one important edition in 22 volumes (Paris, 1819–20). Not until the publication of the Pléiade edition of his *Œuvres complètes* edited by B. Gagnebin and M. Raymond (3 vols., Paris, 1959–64) did a satisfactory modern collection appear. An edition of his correspondence appeared in Paris in twenty volumes, in 1924–33, however, and a larger one, edited by R. A. Leigh, has been in course of publication since 1965. Montesquieu suffered a somewhat similar fate. A three-volume collected edition of his writings made its appearance as early as 1758; and the two decades after 1815, when his influence in France was at its height, saw the publication of several others. But then there was a very long gap until the modern edition by A. Masson (9 vols., Paris, 1950–5) saw the light. An edition of Diderot's *Œuvres complètes*, was produced by J. Assezat and M. Tourneux in 1875–7, and one of his letters appeared in 1955–70 under the editorship of G. Roth and J. Varloot; but until recently Voltaire has stood alone among writers of the Enlightenment in terms of the amount of editorial attention given him.

The books devoted to him have from start to finish been sharply divided in their attitudes and often violently polemical. These sharp contrasts of feeling are clearly visible in the comment on him during his own lifetime and the years immediately after his death. Already he attracted far more attention than any comparable figure in Europe. 'The slightest remains, in verse or prose, which relate to the life of M. de Voltaire, or the history of his works, are, at present, much more fought for than the heroic deeds of Zinghis-Khan', wrote a contemporary commentator.[1] Already he aroused enthusiastic support and bitter antagonism. On the one hand an admirer could claim that he was 'without contradiction, the finest

[1] L. M. Chaudon, *Historical and Critical Memoirs of the Life and Writings of M. de Voltaire* (London, 1786), p. iii.

present that Nature has yet given to man'.[1] Yet on the other he was assailed on a wide variety of counts; because of the factual inaccuracies of much of his historical writing, because of his attacks on the power of rulers and encouragement of revolt against them, because of his tendency to fawn on the powerful when this seemed to his advantage.[2] Above all he was violently attacked for his hostility to religion. He had ridiculed the Bible and the sacraments; he had favoured Muslims and pagans against Christians, Protestants against Catholics; he had dared to urge tolerance of dangerous error in religious matters: these were merely some of the more usual charges against him. Even after his death, alleged one of his opponents, he was still alive and 'works ceaselessly against his Creator who has loaded him with honours, with riches and with talents'.[3] He had, wrote the most balanced and impartial of his eighteenth-century critics, by his mocking of organized religion struck away a psychological prop essential to the ordinary man in his daily life. 'In the reformation of Mr. Voltaire, what remains to encourage the weak, to console the wretched, to curb the wicked, and to serve as a sign of union to all men?'[4] Nor were such charges in any way confined to his native France. To German writers of the *Sturm und Drang* period, and even more to those of early Romanticism, his levity was intensely irritating and often gratefully seized upon as evidence of the essential superficiality and inferiority of the Enlightenment in France. Here, as so often and in so many places, his relentless wit grated upon those of a different cast of mind. In England, where until the last twenty years of his life his reputation had stood high, the later decades of the eighteenth century produced a marked reaction against him to which the Revolution in France added venom. Here again he was faced with the most serious of all the charges levied against him in France—that he had undermined faith and weakened

[1] (T. I. Duvernet), *La Vie de Voltaire* (Geneva, 1786), p. 1. This author appears to have written with Voltaire's approval and in part on the basis of materials supplied by him. See also J. P. L. de la Roche du Maine, *Éloge de Mr. Arouet de Voltaire* (1778); C. Palissot de Montenoy, *Éloge de M. de Voltaire* (1778); J. F. de la Harpe, *Éloge de Voltaire* (1780).

[2] (R. de Bury), *Lettre sur quelques ouvrages de M. de Voltaire* (Amsterdam, 1769), pp. 57–73; (C. F. Nonotte), *Les Erreurs de Voltaire* (Paris, 1767), i. xvi–xvii; S. N. H. Linguet, *A Critical Analysis and Review of all Mr. Voltaire's Works* (London, 1790), p. 3.

[3] *Lettres de feu M. de Haller contre M. de Voltaire* (Berne–Lausanne, 1780), i. xii.

[4] Linguet, *Critical Analysis*, p. 242. Part III of this book is the best contemporary or near-contemporary refutation of Voltaire's attitude to religion.

the churches as bulwarks of social and political stability.[1]
All these accusations were to be repeated with little variation
throughout the nineteenth century. The charge of superficiality,
already made by many of his contemporaries, such as Montesquieu,
Hume, and Frederick II continued to be heard. Thus a widely read
work of the first years of the century attacked his unbelief, which
became more blameworthy as he grew older, and still more his
superficiality and overweening vanity, 'this continual instability,
this lack of reflection, and above all this immense love of success and
fashion'.[2] Carlyle's well-known onslaught some decades later on his
'inborn levity of nature' and 'entire want of earnestness'[3] had had
many predecessors. Under the Restoration attacks on him multiplied.[4] Sometimes they became notably more violent in tone; here the
fears and animosities generated by the Revolution can be seen once
more at work. Thus Paillet-de-Warcy, the most bitter of his critics,
who claimed to have used a mass of documents and to have analysed
and referred to close on 300 of Voltaire's own works, assailed him
with a personal venom hardly equalled in the eighteenth century.
Voltaire's interest in the cases of Calas, Sirven, La Barre, and Lally-
Tollendal, he alleged, stemmed not from any love of justice but
merely from his desire for fame and his hatred of religion. He had
no proper family feeling, was a bad Frenchman and a bad friend,
ungrateful, vindictive, avaricious, and conceited.[5] On the other hand
the great infidel certainly did not lack friends and readers under the
Restoration. His works enjoyed an unprecedented sale in France,[6]
and at least one lengthy work was written with the express purpose of
vindicating his character and personality, 'the beauty of his genius
and the extent of his understanding' against detractors.[7]

[1] There is a detailed treatment of this reaction in B. N. Schilling, *Conservative
England and the Case against Voltaire* (New York, 1950), *passim.*

[2] P. de Brugière de Barante, *De la littérature française pendant le dix-huitième
siècle* (Paris, 1809), pp. 56–8.

[3] *Critical and Miscellaneous Essays* (London, 1899), i. 409–10.

[4] e.g. E. M. J. Lepan, *Vie politique, littéraire et morale de Voltaire* (Paris, 1817);
F. A. J. Mazure, *Vie de Voltaire* (Paris, 1824). Both of these are relatively
moderate.

[5] L. Paillet-de-Warcy, *Histoire de la vie et des ouvrages de Voltaire* (Paris, 1824),
i. 187–9, 410–17.

[6] Twenty-eight complete editions of them appeared in 1815–35.

[7] (J.-L. Wagnière and S.-G. Longchamp), *Mémoires sur Voltaire et sur ses
ouvrages, par Longchamp et Wagnière, ses secrétaires* (Paris, 1826). This book is
really a collection of miscellaneous pieces relating to Voltaire, of which the
memoirs of Longchamp are perhaps the most important.

To detail fully the voluminous nineteenth-century polemics for and against Voltaire would be to weary the reader and waste space. Much of this writing is so repetitive and cast so much in the form of rhetorical assertion rather than any real attempt at analysis that it can be passed over briefly or ignored. It seems clear, however, that the debate over his place in history reached a climax on two occasions: in the decade from the middle 1820s onwards, and again towards the end of the 1870s. More and more, on both occasions, Voltaire tended to be treated as a symbol, a peg on which a general view of the Enlightenment might be hung. Thus he could be seen in the 1850s (very inaccurately) as the representative of the poor and oppressed, an apostle of freedom and tolerance against the forces of despotism and superstition, a great liberating force whose supreme work was a posthumous one, the Revolution which was merely 'the words of Voltaire in arms'.[1] Almost simultaneously there appeared the first scholarly discussion of his place in literature, a favourable one which pointed out in particular that though he had not himself written critical history he had cleared the way for it.[2] Yet the same years saw him denounced as 'the scourge of the eighteenth century and of the beginning of this one', as 'the enemy of his country, insensible to the sufferings of the people and rejoicing in its abasement', and even as 'a second-class writer'.[3] Sometimes this personal denunciation was abandoned for more generalized attacks on the Enlightenment as a whole, attacks for which Voltaire and his works became little more than a pretext.[4] But by far the most serious and sustained onslaught of this period was that made by the Abbé M. U. Maynard in his *Voltaire: sa vie et ses œuvres* (2 vols., Paris, 1867). A Catholic conservative, Maynard loathed the eighteenth century, 'this century of lies'.[5] The book, however, is not a refutation of the ideas of the Enlightenment but a long and generally well-informed

[1] E. Noel, *Voltaire* (Paris, 1855), especially Preface pp. vi–viii, and pp. 331 ff.; A. Houssaye, *Le Roi Voltaire* (Paris, 1858), pp. 374, 377–8, 417.

[2] A. Villemain, *Cours de littérature française: Tableau de la littérature au XVIIIe siècle*, ii (New edn., Paris, 1855), especially pp. 43–52. There was also in the 1850s and 1860s a good deal of rather trivial discussion of Voltaire's literary abilities; e.g. Turpin de Sansay, *Voltaire (sa vie—ses œuvres)* (Paris, 1867), and M. A. Anot, *Études sur Voltaire* (Poitiers, 1864).

[3] M. V. Postel, *Voltaire. Philosophe, citoyen, ami du peuple* (Paris, 1861), pp. 2, 166.

[4] e.g. L. F. Bungener, *Voltaire and his Times* (Edinburgh–London, 1854). This book is a translation of a French original of 1851.

[5] i. 7.

narrative of Voltaire's life, certainly the most complete to appear
hitherto. In it all his faults and weaknesses are detailed with a loving
care bred of deep dislike. He was, Maynard claimed, greedy and unscrupulous where money was concerned (this was one of the first works
on Voltaire to give real attention to his complex financial affairs).[1]
He made great efforts to escape paying taxes on his estate at Ferney,
successfully evaded the payment of postage on his immense correspondence, and had large investments in the Nantes slave trade.
Though he enjoyed for sixty years an average income of over 100,000
livres he left derisory legacies to men who had given him long and
faithful service and was very strict in the enforcement of his own
seigneurial rights.[2] He was vain and vindictive, qualities seen in his
intrigues to achieve membership of the Académie française and
his vicious hostility to critics such as La Baumelle and Fréron.[3]
More serious, he was consumed by hatred of Christianity and an
enemy to the education of workers and the poor. If Calas had been
a Catholic Voltaire would never have lifted a finger to vindicate
him against his judges, while he gave grovelling approval to Catherine
II for her share in the criminal partition of Catholic Poland.[4] As a
personal attack on its subject this book has never been surpassed;
even the reply which it provoked from a supporter of Voltaire was
forced to admit that he 'had not a great character.'[5]

A few years later another burst of polemics was ignited by proposals
for the official celebration of the centenary of Voltaire's death, a
celebration which republicans and radicals in France undoubtedly
hoped to make the pretext for a great anticlerical demonstration.
Once more his critics combined dislike of the Enlightenment, which
he was taken as personifying, with bitter criticism of him as an
individual; and by now it was possible to pillory his writings as the
intellectual seedbed not merely of the Jacobin regime of 1793 but also
of the Commune of 1871.[6] Once more his supporters stressed his

---

[1] The earliest detailed discussion of this subject is probably L. Nicolardot,
Ménage et finances de Voltaire (Paris, 1845), a book bitterly hostile to Voltaire
and to the Enlightenment in general.

[2] i. 106–8, 342–54, ii. 253–8, 464, 625.

[3] ii. Bk. III, chap. 2, Bk. IV, chap. 3, *passim.*

[4] ii. 416–19, 440, 468–83, 486–7.

[5] F. T. Courtat, *Défense de Voltaire contre ses amis et contre ses ennemis* (Paris,
1872), pp. 193 ff., 217.

[6] See J.-A.-P. Dupanloup, Bishop of Orleans, *Premières lettres à Mm. les
membres du conseil municipal de Paris sur le centenaire de Voltaire* (Paris, 1878),

position as an intellectual and political liberator, the symbol of all the highest hopes of France in the last decades of the old regime.[1] And in these uncritical and polemical terms he continued to be discussed until far into the twentieth century.[2]

But the position was slowly changing. Voltaire's readership had always been a middle-class one. His influence had been most pervasive when, as under the Restoration, middle-class objectives such as parliamentary government and freedom of the press seemed threatened by clerical conservatism. From the 1850s onwards, under the threat of socialist revolution, the French middle classes became appreciably less anticlerical. Hostility to organized religion now slowly descended the social scale. From a bourgeois attitude it became increasingly a working-class one; and the working classes had little appetite for the ironies of Voltaire with their markedly élitist overtones. During the second half of the nineteenth century, therefore, he was no longer a central figure in the clash of outlooks in France to the extent that he had been before the 1850s. The result was that in discussions of him scholarship could supplement if not replace polemics. Even before the end of the Second Empire Gustave Desnoiresterres had produced, in the eight volumes of his *Voltaire et la société au XVIIIe siècle* (Paris, 1867–76) an extensive biography which, though generally favourable to its subject, was balanced and impartial. He attempted no sustained discussion of Voltaire's ideas or of his position in the intellectual history of France; but on the events of his life he assembled a hitherto unequalled mass of information. A few years later Moland's edition of the works provided another great tool for future scholarship; and a little later still the first scientific bibliography of Voltaire's immensely numerous writings made its appearance.[3] In not much more than two decades it had become possible to raise discussion of this emotionally charged subject to a new level of accuracy and objectivity. Here as elsewhere in the study of

---

and *Nouvelles lettres . . . sur le centenaire de Voltaire* (Paris, 1878); A. de Kerval, *Voltaire: ses hontes, ses crimes, ses œuvres* (Paris, 1877).

[1] A. Jobez, *La France sous Louis XVI* (Paris, 1877–93), i. 540, ii. 136, 403; for an uncritical English statement of this standpoint see J. Morley, *Voltaire* (London, 1872).

[2] For a typical statement of the radical–republican approving attitude see J. Fabre, *Les Pères de la révolution (de Bayle à Condorcet)* (Paris, 1910), Book V, *passim*. For an essentially conventional conservative attack on Voltaire's élitism and hostility to religion see Brunetière, *Histoire de la littérature française classique*, iii. 468–503.

[3] G. Bengescu, *Voltaire. Bibliographie de ses œuvres* (4 vols., Paris, 1882–90).

the eighteenth century[1] professionalism was now increasingly dominating the scene.

Its victory was not immediate. It was perhaps achieved only in the first years of the twentieth century, with the publication in 1906 of Gustave Lanson's *Voltaire*, which offered the first really balanced and scholarly assessment of his work and influence. Neither his intellectual merits nor his defects were lost on Lanson. He saw that as a historian Voltaire was careless, prejudiced, an amateur; but he also credited him with the great merit of having grasped, as few of his contemporaries did, the fact that real history involved criticism. His treatment of religious subjects, especially of the Jews and the origins of Christianity, was deeply vulgar and distasteful; but this could be explained by the extraordinary *naïveté* and low intellectual standards of much of the conventional piety which he was combating. He had no political theory, no plan of an ideal society; but he genuinely desired reform in France of a realistic and limited kind.[2] The sympathetic but by no means uncritical tone which pervades Lanson's book has been the one most often heard in French writing on Voltaire during the last half-century.[3] In more recent years, however, the situation has been changed by the appearance of the first significant works on the subject written in languages other than French. The *Voltaire und sein Jahrhundert* of the Danish writer and critic Georg Brandès (two vols., Berlin, 1923), impressionistic, exaggerated in its language, and highly favourable to its subject,[4] was in its own day one of the most influential of these. Rather similar in some ways, but more learned and better-written, is the last and best of the biographies to appear, the *Voltaire* of Theodore Besterman (London, 1969). The fruit of years of work and study, based on an unequalled knowledge of the events of its subject's life, this book will not be surpassed for many years within its own terms of reference, if indeed it ever is. It is in every sense a labour of love: the author does not pretend otherwise. 'I have been his lifelong admirer this side of idolatry,' he says of Voltaire, 'I have spent many years in close and critical study of his life and works, I live in his house, work in his

---

[1] See above, pp. 36–7 and below, pp. 222 ff.

[2] pp. 163–8, 171–3, 180 ff.

[3] For example in the elegantly written A. Bellessort, *Essai sur Voltaire* (Paris, 1926), which is a collection of essays on different aspects of his work rather than a connected account.

[4] For example in its description of Voltaire as 'a bundle of nerves charged with electricity, which captivated and enlightened Europe' (i. 5).

library, sleep in his bedroom. It would be absurd for me to pretend to cold impartiality.'[1] The book is not merely the most recent but also the most whole-hearted presentation of Voltaire as the supreme liberator of man and the human intellect, the writer who more than any other enabled human beings to break through irrational and stifling restrictions and realize to the full their own potentialities. 'This catalytic instant in man's long struggles to become himself, for man simply *sapiens* to become *philosophicus*, this movement in the history of humanity is called Voltaire', the author sums up.[2]

Of greater importance in many ways from the standpoint of the professional historian, however, is the study by Peter Gay, *Voltaire's Politics: The Poet as Realist* (Princeton, 1959). This too is avowedly the work of a partisan. 'My book is a defence of the Enlightenment', says Gay flatly. 'I honour what Voltaire honoured and I oppose what he opposed.'[3] In particular he is concerned, like Lanson, but at greater length and with heavier emphasis, to refute the allegation that Voltaire was superficial and unrealistic, lacking in depth of feeling and a grasp of realities. On the contrary, 'the variety of his interests and the shifts in his political opinions sprang not from flightinesss but from an empiricist temper, not from detachment but from a deep engagement with reality'. He was a thinker very much in the constructive liberal tradition, constantly involved with practical politics and far from being an unthinking rationalist. In this he was in tune with the Enlightenment as a whole, since it was at bottom a realistic movement of practical reform and 'most of the philosophes were indefatigable opponents of rationalism, metaphysics, and system-building'. Gay admits that the tone of Voltaire's writings is generalized and abstract, and that this tends to conceal his constant involvement in practical politics; but this tone was merely a disguise assumed to deceive the censors of the old regime.[4] This book marks the highest level hitherto achieved by scholarly sympathy with Voltaire, the most detailed and convincing study of his political ideas yet written by one in agreement with them. Yet the subject is not exhausted. A decade after the appearance of Gay's book Professor I. O. Wade crowned a lifetime of Enlightenment studies by the publication of his *The Intellectual Development of Voltaire* (Princeton, 1969), a formidable study of Voltaire's cultural, religious, and philosophical ideas which in over eight hundred pages takes the

[1] p. 17.
[2] p. 531.
[3] Preface, pp. vii, viii.
[4] pp. 9, 15 ff., 26, 17–18.

subject only as far as the 1740s. This massive work underlines forcibly the growth in scale, and in minuteness and accuracy of detail, visible in the work of the last generation in almost every aspect of writing on the eighteenth century.

The overwhelming extent to which, in the nineteenth century, Voltaire typified the Enlightenment in the minds of most commentators is thrown into relief by the surprisingly small amount of serious attention given, by comparison, to inherently greater and often more interesting figures in the intellectual history of eighteenth-century France. Montesquieu, though occasionally highly praised,[1] aroused no sustained interest. He was not, during his lifetime or afterwards, a controversial writer as Voltaire was. Moreover the moderation of his ideas and the subtleties inherent in them greatly limited their general appeal: only for a relatively short period after 1815 had he much practical significance as a force in French political thinking. Albert Sorel's discussion of him, perhaps the first satisfactory general treatment, did not appear until late in the 1880s.[2] The first attempts at serious accounts of Diderot and his work appeared outside France, in the form of the *Diderot's Leben und Werke* of T. C. F. Rosenkranz (2 vols., Leipzig, 1866) and John Morley's *Diderot and the Encyclopaedists* (London, 1878); and even the most substantial twentieth-century study of him is by an American.[3] Even so towering a figure as Rousseau for long received much less attention than was devoted to Voltaire. In his last years and during the generation which followed his death interest in him was keen and debate on the meaning and value of his work was active. He was attacked as a preacher of the dangerous doctrine of natural religion[4] and praised for the depth and exquisiteness of his sensibilities.[5] A number of personal reminiscences by those who

---

[1] For example by the historian Henri Martin, who rightly claimed that the *Esprit des Lois* had its roots in 'depths to which Voltaire never penetrated' (*Histoire de France*, (4th edn., Paris, 1859), xv 408–9).

[2] A. Sorel, *Montesquieu* (Paris, 1887). For some discussion of the reasons for the paucity of nineteenth-century writing on Montesquieu see P. Barrière, *Un grand provincial* (Bordeaux, 1946), p. 544

[3] A. McC. Wilson, *Diderot, the testing years, 1713–1759* (New York, 1957). The *Diderot* of A. Billy (Paris, 1932) is essentially a semi-popular biography on a big scale rather than a discussion of his ideas.

[4] e.g. N. S. Bergier, *Le Déisme réfuté par lui-même* (5th edn., Paris, 1771).

[5] Comte A. J. de Barruel-Beauvert, *Vie de J.-J. Rousseau* (London, 1789). A spirited reply to the attacks made on Rousseau immediately after his death can be found in Mme Latour de Franqueville, *Jean-Jacques Rousseau vengé* (n.p., 1779).

had known him personally were published.[1] A collected edition of his works in thirty-seven volumes appeared in 1793. His influence on some of the Jacobin leaders, above all on Robespierre himself, is well known and was frequently, in the century which followed, to be a pretext for attacks on him.

During the nineteenth century he continued to be a deeply controversial figure; both his assailants and his protectors grew warm in their arguments and eloquent in their pleadings. Under the Restoration Xavier de Maistre, that greatest of all pessimistic and mystical conservatives, attacked him violently in his most famous work, the *Soirées de Saint Pétersbourg* (Paris, 1821); but this produced a spirited reply in the remarkable *Histoire de la vie et des ouvrages de Jean-Jacques Rousseau* of V. D. de Musset-Pathay, a book which, as a source of information about Rousseau's life, was not to be surpassed for the rest of the century. A generation later, under the influence of the upheaval of 1848–9, for the idealism and utopianism of which Rousseau was widely held responsible, writers so otherwise dissimilar as the liberal Lamartine, the ultramontane Catholic Veuillot, and the anarchist Proudhon joined in attacking him. Yet this evoked, in G. Morin's *Essai sur la vie et le caractère de Jean-Jacques Rousseau* (Paris, 1851), well informed and well argued, perhaps the best book on him hitherto written.[2] However he remained an ambiguous figure in a way that Voltaire had never been. He could be praised as one of the well-springs of the religious revival of the early decades of the century as well as attacked as the enemy of Christianity. He could attract sympathy because of his emphasis on sensibility and the truths of the heart, doctrines which appeared clearly to set him apart from the other great figures of the Enlightenment, and because of the power and brilliance of his style.[3] But he

---

[1] J. Dussaulx, *De mes rapports avec J.-J. Rousseau et de notre correspondance* (Paris, 1798); *Anecdotes of the Last Twelve Years of the Life of J. J. Rousseau originally published in the Journal de Paris by Citizen Corancez* (London, 1798). A general discussion of contemporary and near-contemporary reactions to Rousseau, from a markedly hostile point of view, can be found in C.-A. Fusil, *La Contagion sacrée, ou Jean Jacques Rousseau de 1778 à 1820* (Paris, 1933), chaps. i and ii.

[2] On these mid-century controversies see A. Schinz, *État présent des travaux sur J.-J. Rousseau* (Paris–New York, 1941), pp. 16–29. This book is somewhat too schematic in its belief that there have been regularly occurring waves of interest in Rousseau at thirty-year intervals between *c.* 1760 and *c.* 1940; but it contains much information upon which I have drawn freely.

[3] e.g. C. Estienne, *Essai sur les œuvres de J.-J. Rousseau* (Paris, 1858), p. 25, where the author speaks of 'the harmony and brilliance of magical words which

could also be abused as having rejected reason and objective truth in favour of crude emotionalism and the unreliable subjectivity of his all-devouring 'moi'.[1] Rousseau the romantic, the author of the *Confessions* and of the first part of the *Nouvelle Héloïse*, was not easy to reconcile with the more rationalist author of many of the political and philosophical works; in the nineteenth century this problem tended to be solved, or rather ignored, by a persistent undervaluation of what was realistic and rational in his writings. Moreover Rousseau posed, in a way that Voltaire did not, the problem of the relationship of the writer to his works. Repeatedly the irregularities of his personal life, above all his notorious abandonment of his children to a foundling hospital, were used to attack his works; the personality of the man and the ideas given expression by him were conflated in an often highly misleading way. Thus the need for serious analysis of what he had written could be once more avoided.

More important still, Rousseau appeared a figure of more limited popular appeal than Voltaire and therefore of less practical importance for either good or evil. So good a judge as Villemain had no doubt that the influence of Voltaire was a more real and durable factor in the life of Europe than that of Rousseau, whom he dismissed as 'in that class of speculative writers and eloquent men who simply do not persuade'.[2] Certainly Rousseau, a far more truly revolutionary figure, relatively seldom aroused the really vicious dislike which marks so much conservative comment on Voltaire. His passionate denunciations were easier for many people to stomach than the wounding ironies of the great satirist. At least equally important, he had preached love of country, religious sentiment (though not a specific form of belief), the beauty and importance of family life—all things which conservatives deeply valued. A richer, more complex, and far more ambiguous and self-contradictory thinker than Voltaire, it was easier for him to be all things to all men. It is significant of his failure to arouse the violent passions which Voltaire could

---

overcome us by seducing us'. cf. L. Ducros, *J.-J. Rousseau* (Paris, 1888), pp. 217–18, 222–23.

[1] e.g. the very violent attack by L. I. Moreau, *J.-J. Rousseau et le siècle philosophique* (Paris, 1870), which accuses him of being 'a reasoner who is perpetually unreasonable' and of 'professing contempt for truth' (quoted in Schinz, *État présent*, pp. 35–6).

[2] *Cours de littérature française: Tableau de la littérature au XVIIIᵉ siècle*, ii. 304–5.

still generate that when in 1878 the centenary of the death of each was celebrated in Paris the ceremonies associated with Rousseau were notably less impressive than those devoted to his rival.[1] The same contrast is visible in the fact that though a public monument to him in Paris had been proposed as early as 1790 none was in fact erected until 1889; Voltaire had been given one over two decades earlier.

The early years of the twentieth century saw a large number of attacks on Rousseau at least as bitter as anything hitherto produced. This was in part the product of the resentments, nationalist and Catholic, generated by the outcome of the Dreyfus afiair. Nevertheless the same period also witnessed the beginnings of modern scholarly and impartial work on this fascinating and deeply difficult man. In 1904 the foundation in Geneva of the Société Jean-Jacques Rousseau, which began almost at once to publish a periodical, the *Annales Jean-Jacques Rousseau*, did much to pave the way for the systematic and specialized research which, in this area of eighteenth-century studies as in all others, was to be the hallmark of our own time. Within a few years there began to appear the first weighty and often highly erudite studies of particular aspects of his life and writings; these were increasingly to replace the more wide-ranging, less detailed, and far less scholarly products of the nineteenth century. Thus in 1911 G. Vallette produced, in his *Jean-Jacques Rousseau génévois* the first detailed working-out of an idea which bulks large in much modern discussion of Rousseau— the claim that he was profoundly influenced by the attitudes he imbibed in the Geneva of his youth and that he cannot be understood without some knowledge of the city-state and its life as they were in Rousseau's own day. The dogmatic and oratorical elements in his literary style, his extreme individualism, even his liking for solitary country walks—all these, Vallette argued, were derived from his Genevan background and experience.[2] In Rousseau's career and influence could be seen, he asserted in a not totally unjustifiable access of municipal patriotism, 'Geneva directing for the second time, after three centuries, the world of thought.'[3] A rather different type of specialized study appeared when in 1915 C. E. Vaughan published in Cambridge the two volumes of his *Political Writings of Rousseau*, still the most important work

---

[1] For a description see Schinz, *État présent*, pp. 39–45.
[2] pp. 438–44.
[3] p. 446.

on the subject. Almost simultaneously P.-M. Masson completed the three volumes of his even more significant *La Religion de Jean-Jacques Rousseau* (Paris, 1916). By showing, with an impressive wealth of documentation, that Rousseau had always felt some sympathy for Catholicism (for example that he had lived as a good Catholic in Savoy and then in Paris, and that his return to Protestantism in 1755 had not prevented his remaining on good terms with Catholic priests) Masson struck a great blow at the belief, widespread in the nineteenth century, that he had been in some essential sense antichristian. The same trend towards an increasingly favourable estimate can be seen in what is still the largest biography, the *Jean-Jacques Rousseau* of L. Ducros. The first of its three volumes, which appeared in Paris in 1907, was in many ways highly critical of its subject; the second and third, which were published in 1917 and 1920, present a much less hostile picture.

In recent decades the flow of specialized and impartial studies has become a flood. Attacks on Rousseau from a more or less traditional Catholic point of view have continued to be made, for example by Jacques Maritain and François Mauriac in the 1920s.[1] Polemical and one-sided views have certainly not vanished. He has still sometimes been presented as a mere romantic, even worse a mere sentimentalist, with the rational elements in his thought played down unduly.[2] But the general picture is one of increasingly detailed and accurate knowledge conveyed in books more and more free from the prejudice and emotion which marked so much writing for almost a century and a half after Rousseau's death. His personality, upon which the scholarly published collections of his correspondence throw much new light, in particular has been discussed in greater depth. A preoccupation with it underlies the most recent biography, Jean Guéhenno's *Jean Jacques: Histoire d'une conscience* (Paris, 1962: English translation, London–New York, 1966), which originated with the ideal of correcting factually and

[1] J. Maritain, *Trois reformateurs: Luther, Descartes, Rousseau* (Paris, 1925); F. Mauriac, *Trois grands hommes devant Dieu* (Paris, 1930).

[2] e.g. the substantial discussion in P. Trahard, *Les Maîtres de la sensibilité française au XVIIIe siècle* (Paris, 1932), iii. See also the violent attack in E. Seillière, *Jean-Jacques Rousseau* (Paris, 1921) which begins by condemning him as 'the propagator, often an effective one, of a Christian heresy of a mystical kind' (Avant-propos, p. i) and ends with the hope that 'the mystical elements of this preaching, which are excessively preponderant, may be balanced, without too much delay, by an exertion of firm reason in the heart of our contemporary society' (p. 453).

in a sense rewriting Rousseau's *Confessions*. More recently still the enigmas of his character and psychology have inspired Ronald Grimsley to produce, in his *Jean-Jacques Rousseau: A Study in Self-awareness* (Cardiff, 1961) the most detailed and convincing analysis of its kind. Grimsley sees him as aspiring to 'a mode of existence which allows him simply to be himself as he really is' and 'a positive form of self-expression characterized by a sense of *absolute personal unity*.' This drive towards self-fulfilment, at the same time impressive and pathetic, is to be seen above all perhaps in his 'insistence on the *ideal of self-sufficiency* and a desire to live 'freed from the tutelage of conventional time'.[1] The same author's *Rousseau and the Religious Quest* (Oxford, 1968) also illustrates, from a rather different angle, this tormented and endlessly fascinating personality.

Rousseau's ideas remain, in spite of much patient and intellectually acute discussion, more difficult to sum up clearly than his character. Their rich and fertile confusion, their puzzling but often suggestive contradictions, still give wide scope for disagreement, though that disagreement is more scholarly and better-informed than ever in the past. In particular the central question, that of how far the elements of sentiment and irrationalism in his writings were counterbalanced or outweighed by those of rationality, still produces widely differing answers. Bernard Groethuysen, in a collection of essays on different aspects of Rousseau's ideas, in which he stressed very suggestively the profound conflict in them of values between which it was necessary to choose (sentiment *or* intelligence—virtue *or* knowledge —civilization *or* liberty—patriotism *or* humanity) found himself driven at last to the conclusion that 'for Rousseau the true life is sentiment, the sentiment which absorbs all the faculties of the soul'.[2] Yet almost simultaneously Robert Derathé argued very cogently that the view of Rousseau as a sentimentalist was 'radically false' and that the elements of feeling and of reason in his works do not conflict in more than a purely verbal way since for him 'right reason is conditional on purity of heart'.[3] This conflict of views is still unsettled. Perhaps it never can be finally settled, since Rousseau,

---

[1] pp. 315, 318.

[2] B. Groethuysen, *J.-J. Rousseau. Les Essais XXXVIII* (Paris, 1949) p. 334.

[3] R. Derathé, *Le Rationalisme de J.-J. Rousseau* (Paris, 1948), pp. 7, 167-8. He agrees however that Rousseau set firm bounds to the scope of human reason and did not share the widespread Enlightenment belief in the possibility of progress through the mere extension of knowledge (p. 176).

more than any great writer on politics and society, gives support to the most widely differing views of his real meaning. No body of writing, not even that of Marx, has in the twentieth century provoked more discussion than his. None remains more difficult to sum up simply, to confine within the strait-jacket of any general formula.

But of his importance and inexhaustible appeal there can be no doubt. So far as the sheer quantity of writing on every aspect of his life and thought is concerned he now far outstrips Voltaire; and it is noticeable that whereas until well into the twentieth century important work of this kind was produced almost entirely in French now a great deal appears in other languages, notably in English.[1] The study of Rousseau, like that of Voltaire, has ceased during the last two generations to be a mainly French preserve.

A NEW DEPARTURE OR AN OUTGROWTH OF THE PAST?

In 1872 John Morley asserted that 'The glories of the age of Lewis XIV were the climax of a set of ideas that instantly afterwards lost alike their grace, their usefulness, and the firmness of their hold on the intelligence of men.'[2] This was to deny any intellectual continuity between the Enlightenment and the seventeenth century. A little over seven decades later Paul Hazard proclaimed that 'virtually all the intellectual views and ideas which as a whole were to culminate in the French Revolution had already taken shape even before the reign of Louis XIV had ended', and that 'virtually all those ideas which were called revolutionary round about 1760, or for the matter of that, 1789, were already current as early as 1680'.[3] This was to see the Enlightenment as the continuation and consolidation of an intellectual movement whose roots extended back at least to the seventeenth century if not further. The contrast between the two attitudes reflects a striking change of view.

---

[1] An almost random listing of some works of the last decade or so may make the point as regards both quantity and provenance: J. H. Broome, *Rousseau: A Study of His Thought* (London, 1963); O. Vossler, *Rousseaus Freiheitslehre* (Göttingen, 1963); M. Einaudi, *The Early Rousseau* (Ithaca, 1968); R. D. Masters, *The Political Philosophy of Rousseau* (Princeton, 1968); Judith N. Shklar, *Men and Citizens: A Study of Rousseau's Social Theory* (Cambridge, 1969); C. H. Dobinson, *Jean-Jacques Rousseau. His Thought and its Relevance Today* (London, 1969).

[2] *Voltaire* (London, 1919), p. 26.

[3] P. Hazard, *The European Mind (1680–1715)* (London, 1953), pp. 445–6, Preface, xviii. The first French edition appeared in 1935.

Morley's standpoint was the normal one during the nineteenth century. To the majority of writers the Enlightenment then seemed, for good or evil, a sharp break with the Christian past of Europe. It appeared as a sudden eruption of ideas and attitudes which might be either inspiring and liberating or weakening and destructive, but which were in their essentials certainly new. Questions of origins, whether those of institutions or of ideas, had not greatly preoccupied the writers of the Enlightenment itself; and for a century after the French Revolution it is rare to find much effort to identify the foundations of the intellectual movement which was attacked and defended with such vigour. When they were discussed these foundations were usually thought of—Taine is a good example of this— largely in terms of the influence of the physical sciences and of liberal ideas imported from England. Nevertheless the nineteenth century was an age historical in its modes of thought as none before had been. One of its increasingly dominant assumptions was that any phenonenon in the man-made world, historical, social, or intellectual, could be explained best, and perhaps only, in terms of its origins and development. By the first decades of the twentieth century, therefore, the Enlightenment was being discussed increasingly in terms of what had preceded it; its debt to what had gone before was being more and more investigated and stressed. In France Gustave Lanson and his pupils began to insist that there was no obvious break in intellectual history with the death of Louis XIV.[1] To them the most important achievement of the Enlightenment was to give birth to an *esprit philosophique* which, as it developed, provided the main foundation for modern intellectual and social life. This vision of the Enlightenment as a part of a continuing process rather than a more or less self-contained episode, however important, led them to increasing efforts to push its origins, in the form of rationalism, back into the seventeenth or even the sixteenth century.[2]

A few years later René Hubert, in a book in many ways epoch-making, stressed heavily a quite different intellectual debt of the contributors to the *Encyclopédie*—that to orthodox Catholicism.

[1] Lanson himself spoke of 'a general sliding (*glissement*) of the seventeenth century towards the eighteenth century' (*Revue des cours et conférences*, December 1907, p. 298).
[2] See the comments of I. Wade, *The Intellectual Origins of the French Enlightenment* (Princeton, 1971), p. 16. The first part of this book, 'Enlightenments we have known', is a useful survey of the history of thinking about the nature of the Enlightenment in France.

This, he argued, meant that their thought, far from being truly empirical, was strongly influenced by universalist and *a priori* assumptions. These could be seen in their belief in a human nature which was fixed and unaltering, in their assumption that changes in human societies were essentially the result of variations in the external conditions affecting them, in their faith in natural religion and natural morality, even in the hankering of some of them after a universal language comprehensible to all men. 'Thus the static aspect of things can clearly be seen, in their thought, to gain the upper hand over the dynamic one. This meant an indirect return to an absolutist point of view which was that of the theories against which they were struggling.'[1] The nature of the debt which the Enlightenment owed to the past was thus still very much a matter for discussion. But of the reality of the debt there was now no doubt. The coping-stone may be said to have been placed on this structure of certainty by the publication 1932 of *Die Philosophie der Aufklärung* by the German philosopher Ernst Cassirer, which remains today the most penetrating and satisfactory work on its subject. Cassirer recognized in magisterial terms that 'far more than the men of the epoch were aware, their teachings were dependent on the preceding centuries. Enlightenment philosophy simply fell heir to the heritage of these centuries; it ordered, sifted, developed and clarified this heritage rather than contributed and gave currency to new and original ideas.'[2] The Enlightenment, Cassirer argued, changed the meaning of many of the ideas which it took over from the past, refurbishing them, and transforming them into weapons to be used in the cause of change. In this sense it gave a new form to intellectual activity. But the element of true originality in it was small. This is a rather more extreme statement of the position than many historians would happily accept even today; nevertheless the immensity of the Enlightenment's debt to its predecessors must now be considered as established beyond question.

The precise nature of that debt, however, is not easy to determine. One tempting and still fashionable answer is to stress, as Hubert did, the degree to which the Enlightenment borrowed more or less unconsciously from the Catholic past. This is the theme of the most famous, or at least the most provocative, work written in English

---

[1] R. Hubert, *Les Sciences sociales dans l'encyclopédie* (Lille, 1923), pp. 358.
[2] *The Philosophy of the Enlightenment* (Boston, 1955), Preface, p. vi.

on the intellectual history of the eighteenth century, Carl Becker's
The Heavenly City of the Eighteenth-century Philosophers.[1] Throughout this witty and ultimately unconvincing book the author recurs to
the same theme; the legacy which the *philosophes* unconsciously
took over from the Catholic Middle Ages. 'If we examine the
foundations of their faith', he asserts, 'we find that at every turn the
Philosophes betray their debt to mediaeval thought without being
aware of it.' Faith in nature and reason replaced the authority of
church and Bible; a belief in the redeeming power of grace was replaced by one in the inspiration to be drawn from virtue: but the
change was one of form rather than substance. The Enlightenment
was, in fact, a kind of religion; the *philosophe* version of the progress
of human life provided a kind of echo (though one deprived of its
supernatural elements) of the traditional Christian view of man's
destiny.[2] These claims are vulnerable to criticism from several directions. It must be remembered, in the first place, that the book was
never intended to be more than an extended *jeu d'esprit;* it was not
meant as a complete and rounded picture of its subject and it is
significant that it has never been translated into French and is not
widely known in continental Europe. Even within its own limits,
however, it is unconvincing. Becker showed no adequate grasp of the
fact that the same word may mean very different things at different
times, especially when these times are centuries apart. Both Aquinas
and Voltaire believed in 'reason'; but the term meant different things
to the two men. In the same way 'faith' changed its meaning significantly between the thirteenth and the eighteenth century.[3] Again the
writers of the Enlightenment, whatever their flaws, did with few
exceptions stand for free inquiry and uninhibited discussion; it is
difficult to reconcile this attitude with the medieval attachment to

---

[1] (New Haven, 1932); my references are to the paperback edition of 1959.

[2] pp. 30–1, 48–9, 102, 125 ff. This concept of the Enlightenment as a religious
movement is quite different in both tone and substance from the work of writers
such as Talmon (see above pp. 74–6) who have stressed its dogmatic and
messianic characteristics. Becker's book is inspired by the desire to mock rather
than by fear and pays virtually no attention to the harsher elements of dogmatism,
intolerance, and, potentially at least, historical determinism, which Talmon and
Crocker have identified in much eighteenth-century thought. He had already put
forward the idea of a continuity between the Christian Middle Ages and the
Enlightenment in his *The Declaration of Independence: A Study in the History of
Political Ideas* (New York, 1922), pp. 39, 61.

[3] For this and other criticisms see P. Gay, 'Carl Becker's Heavenly City',
Political Science Quarterly, lxxii (1957), 182–99.

finality and uniformity of belief.[1] In general, moreover, Becker lacks rigour in his impressionistic use of evidence, while that which he produces often provides only a shaky foundation for the conclusions he bases upon it.

It is important also to remember that the opponents of the Enlightenment, and not merely orthodox religious ones, derived powerful arguments and much intellectual ammunition from the past. One of the most remarkable books ever written on any aspect of the history of ideas has shown that the concept of a Great Chain of Being, that is of the living world as composed of a vast number of species differing from each other only by small gradations, so that the chain which they formed was without gaps, was not only still alive but now more widely accepted than ever before. Tracing from Plato and the neo-Platonists, it had become influential in medieval Europe; the age of the Enlightenment saw it at its zenith. 'Next to the word "Nature" ', the author writes, ' "the Great Chain of Being" was the sacred phrase of the eighteenth century, playing a part somewhat analogous to that of the blessed word "evolution" in the late nineteenth.'[2] So long as this general acceptance lasted, however, the victory of the Enlightenment could never be complete; for the Great Chain of Being, by its emphasis on variety and diversity, on the sheer differentness of men from one another, challenged the whole drift of the Enlightenment, which was towards 'the simplification and the standardization of life'. Inherent in this very ancient view of the universe was therefore a profound tendency which can only be described as romantic.[3] The few lines devoted to it here (they can hardly be described even as a summary) do not begin to do justice to Lovejoy's great book. Its relevance in the context of this discussion, however,

---

[1] R. H. Bowen, 'The Heavenly City: A Too-Ingenious Paradox', in R. O. Rockwood (ed.), *Carl Becker's Heavenly City Revisited* (Ithaca, 1958), p. 144. On the other hand R. R. Palmer, in an essay in the same collection entitled 'Thoughts on *The Heavenly City*', has concluded that 'At a high level of generalization, the proposition that the Enlightenment represented a secularization of mediaeval Christianity seems to me perfectly tenable' (p. 126); and another writer has pointed out that the *philosophes* 'could persist in a separation of the goal or meaning of life from the present character of their experience, which was not unlike a more traditional otherworldliness' (C. Frankel, *The Faith of Reason: The Idea of Progress in the French Enlightenment* (New York, 1948), pp. 1–2. The debate on Becker's thesis thus cannot be considered to have resulted in a completely hostile verdict.

[2] A. O. Lovejoy, *The Great Chain of Being: A Study of the History of an Idea* (Cambridge, Mass., 1948), p. 184.

[3] pp. 288, 292–3, 297.

is simply that it demonstrates, with wide and impeccable scholarship, the persistence in the eighteenth century of an intellectual inheritance derived ultimately from the very remote past and profoundly hostile to the Enlightenment.

Many other intellectual debts other than that to the Middle Ages have been proposed for the Enlightenment. Descartes has been frequently put forward as one of the major influences on it; and it has been strongly argued that in so far as there was any generally-held 'world-view' during the eighteenth century it owed as much to him as to Newton.[1] From Descartes, it has been claimed, the thinkers of the eighteenth century derived the idea of 'a set of absolute beliefs upon which enquiry and action must rest, and which are not themselves modifiable by further inquiry'.[2] Similar claims have been pushed even further by A. Vartanian in his interesting though overargued *Diderot and Descartes* (Princeton, 1953). Yet if Descartes contributed to the Enlightenment an element of apriorism and perhaps a certain disdain for history it is arguable that Spinoza contributed something at least equally important: a radical criticism of conventional religious belief and assumptions which sometimes bordered upon out-and-out materialism. The extent, if not the reality, of the contribution is, however, still very much a matter of opinion. Professor Wade, after a study of over a hundred manuscripts of differing degrees of radicalism which were clandestinely circulating in France during the first half of the eighteenth century, concluded that the greatest single intellectual influence which they embodied was that of Spinozist scepticism, and that 'one is tempted to see in the whole movement [the early Enlightenment] a gigantic manifestation of Spinozism triumphant over other forms of thought'.[3] But even he has to admit that there was little real knowledge of Spinoza's ideas in France during this period. Though freely discussed and condemned they were hardly understood in any genuine or meaningful way; and a more recent writer has argued that Wade's estimate of Spinozist influence in the France of the early eighteenth century is greatly exaggerated.[4] The most complete study of the subject, by

[1] e.g. H. Guerlac, 'Newton's Changing Reputation in the Eighteenth Century', in R. O. Rockwood (ed.), *Carl Becker's Heavenly City Revisited*, p. 15.

[2] Frankel, *The Faith of Reason*, p. 154.

[3] I. O. Wade, *The Clandestine Organization and Diffusion of Philosophic Ideas in France from 1700 to 1750* (Princeton, 1938), p. 269; see also Part II, chap, i, *passim*.

[4] J. S. Spink, *French Free-Thought from Gassendi to Voltaire* (London, 1960),

Paul Vernière, concludes judiciously that though the influence of Spinoza in eighteenth-century France was undoubtedly considerable he was none the less a somewhat shadowy figure whose works had been very imperfectly assimilated.[1]

Medieval, Cartesian, Spinozist: if the legacy of the past to the Enlightenment was all these things, it was others as well. It was in one important aspect a legacy from Greece and Rome, in particular from one or two highly admired and influential authors such as Cicero and Horace. The significance of this classical influence has been recently discussed at length by Professor Gay in his *The Enlightenment: An Interpretation. The Rise of Modern Paganism* (London, 1967). Combined with influences derived from the physical sciences, he argues, it underlay the increasing hostility of the *philosophes* to revealed religion and the Christian past. 'The Enlightenment', he declares, 'was a volatile mixture of classicism, impiety and science; the philosophes, in a phrase, were modern pagans'; and he speaks of 'the classical and scientific modes of thought, which converged in the eighteenth century to produce the peculiar amalgam which is the Enlightenment'.[2] This view, even though buttressed by great learning, is clearly incomplete and one-sided. The author himself is careful to point out that by no means the whole of such a highly complex movement of ideas can be described as hostile to Christianity. Such feelings were, he realizes, much more dominant in France than among the *Aufklärer* in Germany, and among a tiny minority of infidels such as Gibbon and Hume in England rather than in the Deist current which was far more widespread and vocal there. But though Christian doctrine, Christian ideals, and theological controversy all remained powerful influences in intellectual life, none the less, he contends, 'religious institutions and religious explanations of events were slowly being displaced from the centre of life to its

---

pp. 238 ff., 280. This book brings out well the strength of the critical, free-thinking, and sometimes materialist currents which ran under the surface of orthodoxy and stability presented by the reign of Louis XIV.

[1] *Spinoza et la pensée française avant la révolution* (2 vols., Paris, 1954). Vernière points out that in the eighteenth century the works of Spinoza were very hard to come by, and that there were not more than a few hundred copies of them available in France. After the appearance of the Latin editions of 1670 and 1677, and the small French translation of 1678, there were no further editions of them at all until 1802 (ii. 696).

[2] pp. 8, 313–14.

periphery'.[1] This is a stimulating book rather than a completely convincing one; but its value as a detailed discussion of one element in the intellectual heritage of the Enlightenment is hardly open to doubt.

Finally, roots of certain aspects of the Enlightenment have been found in the day-to-day political life of Europe during the seventeenth and early eighteenth centuries. The most distinguished of all Italian writers on modern intellectual history, Franco Venturi, has argued that 'the roots of republican thought, from Montesquieu to Rousseau, were deep in a European experience which was recent and not at all mythical'. Many parts of Europe—some of the Italian states, the Dutch Republic, the German cities, Switzerland, England, Poland—contributed, he asserts, to the creation and fostering of a republican tradition which was a living reality in the intellectual life of the Continent during at least the first half of the eighteenth century.[2] The invasion of the Dutch Republic in 1672 by the French, and their bombardment of Genoa twelve years later, he sees as opening an age of struggle in western Europe between absolutist monarchies and old but still viable republics. This was to last until the middle of the following century: the Genoese revolt of 1746 against Austrian rule was 'the last great blaze of this republican and communal tradition'.[3] In England there was a considerable survival of republican influences for several decades after the Restoration of 1660. By the end of the seventeenth century this was becoming fused, in the work of such writers as Molesworth and Toland, with radical religious thought: in the form of deism, of pantheism, perhaps even of Freemasonry, this amalgam was to become an important influence in the

[1] pp. 326, 331, 338. The best discussion in English of the continuing importance of traditional Christianity and its modes of thought had been provided a generation earlier by R. R. Palmer in his *Catholics and Unbelievers in Eighteenth-Century France* (Princeton, 1939). This careful study of Jesuit and Jansenist thought is a valuable corrective to any easy assumption that the *philosophes* were the only writers of their age who had a view of man and society which deserves serious consideration. Perhaps I should add here that I have deliberately avoided, for reasons of space, any discussion in the preceding paragraphs of specific debts to the past owed by individual writers or thinkers of the eighteenth century. There are many discussions of these: the best as well as the best-known is J. Derathé *Jean-Jacques Rousseau et la science politique de son temps* (Paris, 1950), the conclusions of which are conveniently summarized in A. B. Cobban, 'New Light on the Political Thought of Rousseau', *Political Science Quarterly*, lxvi (1951), 272–84.

[2] F. Venturi, *Utopia and Reform in the Enlightenment* (Cambridge, 1971), pp. 18 ff.

[3] pp. 23–38.

intellectual life of continental Europe.[1] Shaftesbury and Diderot in particular acted as channels through which it was transmitted; to it the vague but widespread and potent ideal of republican virtue, and the sense of political duty and civic pride associated with it, owed much. Like all the discussions of Enlightenment origins considered here this one is incomplete: it does not pretend to be otherwise. But it illuminates, to a greater extent than the work of Becker, Gay, or Vernière, an influence, hitherto relatively neglected, which had real importance in the practical politics as distinct from the intellectual constructs of the eighteenth century.

THE TRIUMPH OF THE BOURGEOIS SPIRIT?

The eighteenth century was well aware that there was often a relationship between the play of economic forces on the one hand and the distribution of political power on the other. To such an acute observer of the situation in France as the Abbé Galiani, for example, himself a would-be reformer in his own state of Naples, it was clear that political or social programmes were a reflection of, among other things, the economic interests of those who advocated them.[2] Such insights foreshadow the belief, which was to become important in our own century, that the Enlightenment was at bottom an expression of middle-class interests, the call of an increasingly powerful, educated, and self-confident bourgeoisie for the liberty and opportunity which the *ancien régime* continued to deny it. In an organized and dogmatic form such a belief is above all the work of Marxism. Not merely Marx himself but the followers who expounded and elaborated his ideas—Engels, Lenin, Plekhanov—thought of the *philosophes* as middle-class revolutionaries, or at least as having prepared the way for a middle-class revolution.

We know today [wrote Engels of the Enlightenment in 1878 in his *Anti-Dühring*] that this kingdom of reason was nothing more than the idealized kingdom of the bourgeoisie; that this eternal Right found its realization in bourgeois justice; that this equality reduced itself to bourgeois equality

---

[1] Chap. ii, *passim*. Republican survivals in eighteenth-century England are illustrated in detail in Caroline Robbins, *The Eighteenth-century Commonwealthman* (Cambridge, Mass., 1959) and B. Bailyn, *The Ideological Origins of the American Revolution* (Cambridge, Mass., 1967).

[2] See the discussion of his ideas, stressing their realism and materialism, in H. Holldack, 'Der Physiocratismus und die absolute Monarchie', *Historische Zeitschrift*, cxlv (1932), pp. 520–7.

before the law; that bourgeois property was proclaimed as one of the essential rights of man; and that the government of reason, the Contrat Social of Rousseau, came into being, and only could come into being, as a democratic bourgeois republic.[1]

In particular the more materialist French thinkers of the eighteenth century—Diderot, Holbach, Helvétius, La Mettrie—aroused admiration; Marx himself saw them as one source of his beliefs. Their materialism, admittedly, was incomplete and imperfect. It was contaminated by elements of idealism, since it expected fundamental change in France to come through the work of a reforming monarch. It was excessively and unimaginatively mechanistic, because of the backwardness of the biological sciences in the eighteenth century by comparison with physics and mechanics. Yet by the standards of its own age it was advanced and progressive.[2]

This attitude has dominated all Soviet discussion of the Enlightenment and its significance. In this discussion such terms as 'the Age of the Enlightenment' have been sternly rejected; they imply much too strongly an idealist and non-materialist view of the dynamics of history. What is significant in the Soviet view is the paving of the way for the French Revolution, a process in which the Enlightenment played a part, rather than the Enlightenment for its own sake. Its importance, in other words, lies not in the quality of its ideas or any novelty in its insights, but in the fact that it was a great attack on feudal survivals launched by a growing and progressive middle class. It was therefore an essential stage in the historically inevitable unfolding of capitalism. Neither Voltaire, witty, sceptical, and fundamentally uninterested in political questions, nor Rousseau, a passionate and confused idealist, fits well into this scheme of values. Hume's profoundly destructive analysis of prevailing intellectual assumptions, or Herder's stress on the profound differentness of national groups from one another and the right of each to work out its own destiny, agree with it no better. It is above all Diderot who has seemed to Soviet eyes the most praiseworthy and the most distinctly modern of the great intellectual figures of the eighteenth century. His materialism; his atheism; his alleged anticipation of dialectical modes of thought; his belief in the capacity of science and technology

[1] English edition (Moscow, 1954), p. 29.

[2] A. Miller, *The Annexation of a Philosophe: Diderot in Soviet Criticism* (Diderot Studies, XV: Geneva, 1971), pp. 25–32. Though pedestrian in its style and presentation, this book is the best collection in English of detailed information on the Soviet view of the Enlightenment.

to transform man's environment (seen in the marked interest in technology of the *Encyclopédie*): all these have earned him in Soviet historiography a position higher than that which he normally occupies in western writing on the Enlightenment. His position in the history of literature as a precursor of 'critical realism'; his generally realist attitude to aesthetics; even his admiration for Russia, which alone among major *philosophes* he visited, and for which he predicted a brilliant future: these have raised his status still further in the eyes of orthodox Marxists. More than any other *philosophe*, he can be seen as 'a precocious traveller on the road that led to Marx and Engels and onward to the Soviet State'.[1]

Not all Marxist or Marxist-influenced writers have taken quite so high a view of Diderot. Morelly, whose *Code de la Nature* (1755), with its proposals for the abolition of private property and for the establishment of a system of national assistance for the poor and of the right to work, is perhaps the most obviously communist book produced by the eighteenth century, has also attracted sympathetic attention.[2] But his socialism was purely utopian. It was not rooted in any effort at economic analysis or concept of class struggle and could be made to fit any political regime. Economically it was null, indeed retrogade; politically it was conservative, at least by implication. A much more promising candidate for the attention of writers sympathetic to a materialist interpretation of history is that strange, isolated, and still little-known figure, S. N. H. Linguet. His socialism was the result of an analysis of everyday realities: alone of the writers of the Enlightenment he had grasped, albeit in an imprecise and incomplete form, the idea of class struggle. One result of the growing influence in recent decades of attitudes influenced by Marxism has been to raise his status from that of an interesting fringe writer to that of a forerunner of one of the greatest developments of nineteenth-century intellectual life.[3]

Belief in the class implications and affiliations of the Enlightenment and its work has attracted many historians who were not Marxists in any complete or dogmatic sense of the term. Such ideas have been

---

[1] Miller, *The Annexation of a Philosophe*, pp. 126, 422.

[2] The most important effort at revaluation is R. N. Coe, *Morelly, Ein Rationalist auf dem Wege zum Sozialismus* (Berlin, 1961).

[3] See the comments on Morelly and Linguet of S. Stelling-Michaud, 'Lumières et politique', *Studies on Voltaire and the Eighteenth Century*, xxvii (Geneva, 1963) 1537–9.

accepted by radicals of a wide variety of types, above all perhaps in the English-speaking world. Writers with little in common save generally critical and reformist attitudes to social and political problems have joined in believing that the hostility of the *philosophes* to the church, their stress on the freedom of the individual, and their belief in the importance of rational utility as a guide to policy, were at bottom a response to the needs of a middle class resentful of traditional restrictions.[1] The same assumptions have often coloured responses to individual writers. Thus it was possible, to take only one of many examples, for a leading English radical of the 1930s to speak of Voltaire as a 'thinker to the middle-class revolution' whose task it was to 'give confidence and direction to the middle class'.[2] In continental Europe, however, discussion of the Enlightenment in terms of class interests and relations has been in general dominated by Marxism; and here it has, quite logically, taken the form of condemnation of some writers as well as of praise of others. Montesquieu, the greatest of all modern conservatives, realistic and profound, is an obvious target. He has been frequently presented as the representative of the interests of the nobility in France, above all of those of his own section of it, the higher ranks of the *noblesse de la robe*. His defence in the later chapters of the *Esprit des Lois* of noble privileges and seigneurial rights can be used to support these charges. Albert Mathiez, the most notable Marxist among French historians of the first decades of this century, went so far as to speak of 'this red reactionary called Montesquieu', and to allege that the *Esprit des Lois* 'is not a progressive book; it is a retrograde book'.[3] Rousseau is so completely an individual that it is difficult to fit him into any framework based on economic interests or social classes. Even here, however, the attempt has been made. One of the

[1] A good example is to be found in the work of the best-known English radical of the inter-war period; H. J. Laski, *The Rise of European Liberalism: An Essay in Interpretation* (London, 1936), pp. 168–70, 172–5. The same ideas are put forward in Laski's article 'The Rise of Liberalism', in *Encyclopedia of the Social Sciences*, i (London, 1930). For a comparable French example see H. Sée, *L'Évolution de la pensée politique en France au XVIIIe siècle* (Paris, 1925), pp. 360–3.

[2] H. N. Brailsford, *Voltaire* (London, 1935), pp. 34, 114. It should be added, however, that this attitude is by no means obtrusive in the book, which is essentially a competent though laudatory biography.

[3] A. Mathiez, 'La Place de Montesquieu dans l'histoire des doctrines politiques du XVIIIe siècle', *Annales historiques de la révolution française*, n.s.,vii (1930), 109, 110. A similar denunciation can be found in Mathiez's violently hostile review in the same publication (n.s., iv (1927), 509–13) of E. Carcassonne, *Montesquieu et le problème de la constitution française au XVIIIe siècle* (Paris, 1927).

most extreme recent discussions of this kind sees him as, at least from the 1760s onwards, providing the French aristocracy (how consciously is not made clear) with an ideology which could be used to support its view of life and endow it with renewed self-confidence. In particular, it is argued, the *Nouvelle Héloïse*, much his most popular book in his own lifetime, served this purpose in its call for an autarchic, paternalistic, agricultural society in which luxuries were unknown. Though Rousseau, the author admits, denounces the existence of privileges based merely on birth, these denunciations are never total and never rise to a complete rejection of all social conventions.[1] The argument, however interesting, presents a curiously narrow view of one of the greatest figures in intellectual history. It also demonstrates, none the less, how even a man so endlessly discussed can still be seen from a new angle and in an un unexpected light.

Ultimately, however, any explanation of the Enlightenment in terms of classes and their economic interests seems implausible. Many of the *philosophes*, after all, were members of the privileged orders of church and nobility; and these included some of the most radical, such as Condorcet and Holbach. More important still, in so far as the middle class played an active role in the Enlightenment at all in France it was as a professional class of doctors, scientists, technicians, and lesser officials rather than as one of merchants and industrialists. Much of the *Encyclopédie* was written by middle-class men, as René Hubert first showed convincingly over half a century ago; but they were professional men with a liberal, gallican, and monarchist outlook, quite lacking in desire to undermine the foundations of society.[2] In general it was the most radical works of the period which were slowest to reach a substantial public. Finally, if some process of cause and effect linked economic progress with the production of radical ideas, how are we to explain the fact that England, on the

---

[1] J. Biou, 'Le Rousseauisme, idéologie de substitution', in *Roman et lumières au XVIIIe siècle* (Paris, 1970), pp. 115–27.

[2] R. Hubert, *Les Sciences sociales dans l'encyclopédie*, pp. 15–22. It should be noted, however, that Stelling-Michaud, in the best short presentation of a moderate Marxist view of the Enlightenment, argues that this monarchism, whose existence he does not deny, was merely a defence of middle-class interests. This was because the monarchy in France seemed to stand for centralization against a backward-looking regionalism and political pluralism which denied the middle-class economic and social opportunity. Middle-class monarchism was thus opportunist rather than a genuine expression of belief ('Lumières et politique', 1530). This view would be generally accepted by Marxist writers.

brink of revolutionary economic change and with the most developed middle class of all the great European countries, produced so little intellectual radicalism during this period? 'That the very country which was moving towards the industrial revolution should be the only one in which the organization of the Enlightenment did not exist, should suffice in itself to call in question the oft-repeated Marxist interpretation of the Enlightenment as the ideology of the Bourgeoisie' is the common-sense view of perhaps the most satisfying writer of the last generation on the subject.[1] The one great eighteenth-century Englishman who can be regarded as truly representative of the European Enlightenment, the historian Edward Gibbon, remained throughout his life an isolated and untypical figure in his own country. True radicalism can be said to make its appearance in eighteenth-century England with Price, Priestley, and Bentham. But none of these became really important figures until the 1790s; and the first two in any case confined their radicalism to demands for religious toleration and parliamentary reform, without paying much attention to social problems. We are driven to accept Laski's conclusion that 'the outstanding feature of English political thought in the eighteenth century is the absence of any original note'.[2] But if political ideas were so stagnant in a society pregnant with the greatest economic transformation in history, what becomes of the belief that they are merely an outgrowth of economic forces and relationships?

THE STREAK OF IRRATIONALISM

Until the last two or three decades the great bulk of discussion of the Enlightenment proceeded on two assumptions. It was believed in the first place that it had been, in spite of all its complexities, enough of a unity to allow of valid generalizations. It was also believed that it had been a movement towards greater rationalism in the ordering of human affairs. In recents years both these assumptions have been challenged with increasing sharpness and insistence. That the Enlightenment was the creation of a small intellectual élite, that it had no direct impact whatever on the ordinary man, had always been recognized, even if not always explicitly admitted, by intelligent commentators. That from the 1760s onwards there had been a

---

[1] F. Venturi, *Utopia and Reform*, p. 132.
[2] *The Rise of European Liberalism*, p. 210.

growing cult throughout much of Europe of feeling and sensibility, a growing revulsion against classical models felt to be exhausted and emotionally unsatisfying, had, also been recognised by scholars since at least the 1820s.[1] Rousseau's *Nouvelle Héloïse* (1762), Bernardin de Saint-Pierre's immensely influential novel *Paul et Virginie* (1788), Goethe's even more influential *Werther* (1774), could all be seen clearly as merely the crest of a wave of sentiment, a foreshadowing of romanticism, which had swept across the Continent in these decades. The charge against the Enlightenment, however, is not that it drew to its end in this atmosphere. It is rather that even in France, and even among the educated, it was never a movement simply or perhaps even predominantly of optimistic rationality. On the contrary, it is argued, there were always latent within it deep-rooted elements of pessimism and irrationality; and these, especially the latter, became more prominent as the century drew to a close.

It had always been clear that the Enlightenment had ended, in the 1770s and 1780s, to the accompaniment of a marked growth throughout much of Europe of exotic and sometimes extreme forms of irrationalism. F. A. Mesmer, with his claims to be able to effect marvellous cures by utilizing the powers of the 'universal fluid', which permeated and bound together the whole natural world, enjoyed immense success and popularity in Paris in the early 1780s. So did for a time Cagliostro, the adventurer who claimed to be a thousand years old, to have access to the mystic lore of the ancient Egyptians and to be able to transmute base metals into gold. Simultaneously L. C. de Saint-Martin, 'the unknown philosopher', was preaching an almost unintelligible but emotionally satisfying mysticism which, with its rejection of the rational, the specific, and the measurable, had great appeal in many parts of Europe. None of these men could be considered part of the Enlightenment; but must not there be something suspect about a great intellectual movement which culminated in this way? Certainly there could be no doubt of the strength of irrational and anti-rational feeling in the years before the French Revolution. 'Never, certainly', wrote one mesmerist in 1788, 'were Rosicrucians, alchemists, prophets, and everything related to them so numerous and so influential. Conversation turns almost entirely upon these matters; they fill everyone's thoughts; they strike everyone's imagination . . . Looking around us, we see

---

[1] See the list of comments illustrating this in A. Monglond, *Le Préromantisme français* (Grenoble, 1930), i. Avant-propos, vi ff.

only sorcerers, initiates, necromancers, and prophets. Everyone has his own, on whom he counts.'[1] Mesmerists, Rosicrucians, Sweden-borgians, alchemists, cabalists, theosophists, mystical forms of freemasonry; the flight from rationalism took a bewildering variety of forms. It was none the less real and widespread. How had a movement allegedly rational and liberating ended in this efflorescence of superstition?

Few writers of the nineteenth and early twentieth centuries faced this question squarely. Soulavie, though he gave a good deal of attention to Mesmer and Cagliostro, made no attempt to explain their vogue and contented himself with the vague remark that 'the eve of the revolution was marked by the most singular reveries of the imagination'.[2] Droz nearly four decades later attributed the new intellectual climate of the years before 1789 merely to an endemic taste for novelty which had suddenly become more marked than usual in Paris.[3] Louis Blanc in the 1840s produced a more thoughtful and rather more convincing explanation. There had been a growth of mysticism, he argued, as an inevitable and indeed healthy reaction against an Enlightenment which had been too analytical, which had sacrificed feeling to reason, the comfort of belief to the mere pride of knowledge. The new messiahs at least preached the unity of the world and rescued the ordinary man from the isolation in which the Enlightenment had left him. 'Saint-Martin glorified the attraction of souls, love; Mesmer the attraction of bodies, magnetism. . . . [T]ogether they proclaimed, dividing between them the two great aspects of life, the dogma of solidarity.'[4] No nineteenth-century writer, however, attempted a proper scholarly analysis of the remarkable growth of mystical tendencies which the last decades of the eighteenth century had witnessed. On freemasonry, a major aspect of this mysticism, a vast controversial literature, almost all of it of low intellectual quality, rapidly grew up. As early as 1792 the idea that the Revolution in France had been the result of a masonic plot began to take shape; from that date onwards it has never ceased to be put forward, often with passion but seldom in works with much claim to scholarly

---

[1] R. Darnton, *Mesmerism and the End of the Enlightenment in France* (Cambridge, Mass., 1968), pp. 70–1. This brilliant little book is the only good account of a fascinating subject.

[2] J. L. Soulavie, *Historical and Political Memoirs of the Reign of Lewis XVI* (London, 1802), vi. 253.

[3] F. X. J. Droz, *Histoire du règne de Louis XVI* (Paris, 1839–42), i. 417 ff.

[4] *Histoire de la révolution française*, ii (Paris, 1847). 72. 108.

respectability.[1] This writing, however, is part of the undergrowth of history: though it is in many ways of great interest a discussion of it is hardly in place here. Against the great outpouring of works on Voltaire and the other leaders of the Enlightenment which marked the nineteenth century there is little comparable discussion of what may be called the anti-Enlightenment to be set.[2] Even as late as 1913 the author of a detailed study of the diffusion of enlightened ideas in the forty years before the Revolution could find no explanation for the decay of the *philosophe* outlook other than the fact that it had already achieved such great practical success (in, for example, the ministry of Turgot in 1770–4). 'The temporal triumph of a doctrine', he wrote, 'has always as an inevitable result its spiritual weakening.'[3]

The work of the last three or four decades, however, has placed it beyond doubt that a deep-rooted strain of irrationalism, of mysticism, often of pessimism, a taste for mystery and the wonderful, a deep-seated wish to believe, flourished throughout the entire course of the eighteenth-century Enlightenment. This was unmistakably so even at its centre, in the educated society of France. Any idea of the intellectual life of the period as completely dominated by reason and rationalist criticism must now be abandoned. 'In this century of reason people never ceased to delight in superstition.'[4] The eighteenth century often tried to rationalize the supernatural; but it never disregarded or despised it. The proliferation of mystical sects and creeds with which the Enlightenment ended was the culmination, not the negation, of what had gone before.[5] The physical sciences, far from being the intellectually aseptic influence, purged of superstition and all the grosser emotions, which historians in the nineteenth century had imagined them to be, were in fact during the Enlightenment still associated with astrology and alchemy in the minds even of many advanced thinkers. Thus in 1711 the Comte de Boulain-

---

[1] J. M. Roberts, 'The Origins of a Mythology: Freemasons, Protestants and the French Revolution', *Bulletin of the Institute of Historical Research*, xliv (May 1971); and the same author's extended study, *The Mythology of the Secret Societies* (London, 1972).

[2] The most important exception to this statement is A. J. Matter, *Saint-Martin, le philosophe inconnu* (Paris, 1862).

[3] J.-P. Belin, *Le Mouvement philosophique de 1748 à 1789* (Paris, 1913), pp. 368–9.

[4] L. Trenard, *Lyon de l'encyclopédie au préromantisme* (Paris, 1958), i. 175.

[5] See the comments of R. Mauzi, *L'Idée du bonheur dans la littérature et la pensée françaises au XVIIIe siècle* (Paris, 1960), p. 12.

villiers, the most radical figure of the early Enlightenment in religious matters, attempted in his *Astrologie mondiale* to draw up parallel tables of historical events and astrological conjunctions as a means of proving the dominance of the latter over the former. A decade later the terrible plague which ravaged Provence in 1720 could still be attributed by educated men to the influence of malign stars.[1] At the end of this period mesmerism fitted very well the view of the universe as pervaded by immensely powerful and deeply mysterious forces which science had made widespread. The discovery of gravitation; the increasing attention paid to electricity and magnetism; the astonishing revelation that air and water, hitherto thought of as elements, were in fact complex substances; all these made the world seem strange and wonderful at least as much as they rendered it more rational and comprehensible. The advancing progress of science generated an increasing credulity as to what was now possible, an increasing belief in miraculous cures, an increasing putting forward of visionary systems to explain the workings of the universe. In this atmosphere it was easy for a mesmerist to believe that 'above science is magic, because magic follows it, not as an effect, but as its perfection'.[2] On a different and more important level of intellectual life, an American historian has argued powerfully that the entire thought of the Enlightenment was deeply influenced by a powerful strain of scepticism, relativism, and distrust of the idea of progress over time. Historical pessimism, in his view, 'had its roots deep in the "philosophical" movement itself'.[3] Far from thinking of man as naturally good, he argues, most *philosophes* saw him as merely neutral, almost a cipher where morality was concerned, clay to be moulded for good or evil by his environment, by education or by mere chance.[4]

[1] J. Ehrard, *L'Idée de la nature en France dans la première moitié du XVIIIe siècle* (Paris, 1963), pp. 42–3.

[2] Darnton, *Mesmerism*, p. 38. Chapter i of this book gives an excellent account of this aspect of the intellectual life of the later 1770s and 1780s in France. A. Viatte, *Les Sources occultes du romantisme: illuminisme-théosophie, 1770–1820* (Paris, 1928), vol. i, which is still the best-known work on late eighteenth-century mysticism, ignores its scientific background.

[3] H. Vyverberg, *Historical Pessimism in the French Enlightenment* (Cambridge, Mass., 1958), p. 1.

[4] The change in attitudes over the last generation can be illustrated by comparing the article 'Enlightenment' in the *Encyclopedia of Philosophy* (New York–London, 1967), with that on 'Illuminismo' in the *Enciclopedia italiana* (Milan, 1933). The latter, though in some ways the more penetrating of the two, stresses that the basic characteristic of the movement was 'the absolute, dogmatic, so to speak religious faith in the unity and validity of human reason'. The former takes

WIDENING VIEWS

The development and popularity of Marxist or Marxist-influenced views of the Enlightenment, and the increasing attention given to the irrational and pessimistic elements in it, are aspects of a widening of intellectual horizons which has more and more characterized the writing of the last generation. This has also involved a widening of geographical perspective. Until within the last two or three decades it was still possible to think of the Enlightenment as an almost entirely French movement. The fundamental importance to it of the work of Locke and Newton, or more indirectly of Bacon, could hardly be questioned. It was usually agreed that in large measure England originated 'enlightened' ideas which were then elaborated and spread by French writers. But though its foundations were largely English the Enlightenment was seen as overwhelmingly French. Sometimes this attitude was little more than an expression of national pride, as in Victor Cousin's assertion that 'the French people is the historic people of the eighteenth century; its character is precisely that of this century'.[1] But whatever its origins it meant that the significance in the Enlightenment of Germany and Italy was generally and often grossly underestimated. Recent writing has increasingly abandoned this restrictive view. France's position as the centre of the Enlightenment, the country in which its questions, its doubts, and its contradictions were sharpest, is beyond challenge. But there is now a clearer realization than ever before that, quite apart from the unique and towering figure of Hume across the English Channel, there were thinkers and writers of great originality and importance elsewhere in Europe.

Two of these above all have benefited from this reassessment—J. G. von Herder in Germany and Giambattista Vico in Italy. Herder has had to wait long for the recognition which he has now achieved. His most important work, the *Ideen zur Philosophie der Geschichte der Menscheit* (1784–91), was translated into English in 1800. A

---

a more open and sophisticated view of the subject, stressing the persistence during the eighteenth century of religious influences hostile to the Enlightenment, the way in which its nature and influence varied in different parts of Europe, and the fact that its thought was not abstract and unrealistic and did not depend on a naïve belief in the natural goodness and/or reasonableness of man.

[1] *Cours de philosophie* (Brussels, 1840), i. 29. Cousin also shows a notable desire to play down the importance of the English roots of many aspects of the Enlightenment.

second English edition appeared in 1803 and a French one (translated from the English) in 1827. But there is no evidence that he had any real influence on English or French intellectual life during the nineteenth century. Even in German the first important study of him, the *Herder, nach seinem Leben und seinen Werken* of Rudolf Haym appeared only in 1880, while the first scholarly complete edition of his works, by B. Suphan, was completed only in 1913. The last generation, however, has seen, notably in the English-speaking countries, a steady stream of books on him and his ideas.[1] More and more he stands out as one of the most truly original intellectual figures of the eighteenth century. Inevitably a good many of his ideas and attitudes can be traced to earlier writers;[2] and the width of his views and feelings (his admiration, for example, for both the Jews and the Slav peoples), puts him in tune with some of the cosmopolitanism of the Enlightenment. But it is difficult to think of any writer of the second half of the eighteenth century, with the exception of Rousseau, who showed such originality as Herder. His view of history, his belief that a man must be of his own age and nation and that every historical moment and situation is unique and different from all others, sets him apart from the main current of the Enlightenment.[3] So do his love of all that is natural and deep-rooted, his deep dislike for uniformity and the distrust of abstract reasoning which pervades his work. His passionate belief in the supreme importance of language, in which he believed the 'whole heart and soul' of a nation were to be found, his belief in communication as in itself a kind of creation, derive not from France or the Enlightenment but from the strange and enigmatic German writer Johann Georg Hamann, 'the Magus of the North', a figure who cannot be neatly categorized in any way.[4] Perhaps most important of all, by making the happiness of the individual and the full development of his potentialities depend absolutely on his membership of a national community, Herder built the bridge which leads from the Enlightenment to the most

[1] e.g. W. Rasch, *Herder* (Halle, 1938); F. McEachran, *The Life and Philosophy of J. G. Herder* (Oxford, 1939); M. Rouché, *La Philosophie de l'histoire de Herder* (Paris, 1940); A. Gillies, *Herder* (Oxford, 1945); R. T. Clark, *Herder* (Berkeley–Los Angeles, 1955); G. A. Wells, *Herder and After* (The Hague, 1959).

[2] I. Berlin, 'Herder and the Enlightenment', in *Aspects of the Eighteenth Century*, ed. E. R. Wassermann (Baltimore–London, 1965), pp. 49–53.

[3] See A. O. Lovejoy, 'Herder and the Enlightenment Philosophy of History', *Essays in the History of Ideas* (Baltimore, 1948), pp. 166–82.

[4] Berlin, 'Herder', pp. 64–5. There is a good short account of Hamann's ideas in R. Pascal, *The German Sturm und Drang* (Manchester, 1953), chap. i.

powerful political beliefs and emotions of nineteenth-century Europe. His position as a thinker in some ways part of the Enlightenment but in others quite alien to it means that the attention given him, as well as being amply justified in its own right, has brought home to scholars in a way never achieved in the nineteenth century the fecundity and variety of eighteenth-century intellectual life. His work is one of the greatest barriers to any attempt to force the thought of the period into a purely French mould.

The same is true of Vico. He, like Herder, was slow to achieve recognition of his rightful stature. He influenced some of the Neapolitan reformers of the later eighteenth century, notably Gaetano Filangieri; and the republication of his works in Italy was begun in 1818. But though Herder had some knowledge of his writings he was almost completely ignored outside the peninsula until the later 1820s, when Michelet began to make him known both by writing about him and by translating some of his work. (A German translation of 1822 had aroused no interest.) During the nineteenth century his fame slowly spread; it was an age to which both Vico's romanticism and his critical philosophy were likely to appeal. Yet in 1911 the most famous of all his advocates complained that 'even today, though well known in certain restricted circles, he has not taken the place he deserves in the general history of thought'.[1] Like Herder he achieves greatness because of a quality of penetrative and creative imagination generally lacking in much of the thinking of the Enlightenment. Like Herder, again, his mind was profoundly historical in cast: he has been described as 'one of the most historically-minded men who ever lived'.[2] One of the basic ideas of his greatest work, the *Scienza Nuova* (first published in 1725 and virtually rewritten in a seond edition of 1730; the final version was published in 1744, the year of Vico's death) is that the true history of humanity is the history of its mental states. But these are to be grasped above all through an immense and testing effort of the imagination, a struggle by the historian to project himself into, and comprehend, an intellectual environment quite foreign to his everyday life. The history of an age is to be understood through the study of its poetry and its use of language, through the interpretation of its mythology as well as through documents and inscriptions. This is an attitude which, as Croce pointed out, fore-

[1] B. Croce, *La filosofia di Giambattista Vico* (Bari, 1911), p. 242.
[2] H. P. Adams, *The Life and Writings of Giambattista Vico* (London, 1935), p. 164.

shadows nineteenth-century developments and breaks away from the narrow and amateurish view of history so widespread during the Enlightenment.[1] More than any other thinker Vico can be called the founder of the philosophy of history.

On the English-speaking world the impact of this confused and difficult but immensely fertile and suggestive writer came relatively late. The first significant work on him in English did not appear until 1884.[2] The last generation or more, however, has seen, as with Herder, the appearance of a steady and even copious stream of work on Vico, much of it of high quality.[3] More than Herder, he is *sui generis*, standing apart from the Enlightenment in France and having no contact with it. Though he lived until 1744 this poverty-stricken teacher, largely self-taught and isolated in remote Naples, had almost nothing in common with French contemporaries such as Montesquieu, Voltaire, Fontenelle, or the young Diderot. His grasp of the constructive power of religion as a force in human affairs in itself sets him far apart from many of the best-known figures in French intellectual life.

Vico and Herder are the supreme illustrations of the fact that France, though the leader of eighteenth-century Europe in the generation and spread of ideas, did not monopolize the Continent's intellectual life. A full realization of this fact, an adequate understanding of the importance of Germany and particularly Italy in this respect, has been slow in coming. It is now visible, however, in a number of ways. The Enlightenment in Russia, in so far as it existed at all, can now be seen clearly as German rather than French in inspiration. Its roots lay in Germany, and particularly in some of the Protestant universities influenced by Pietism—Halle, Marburg, Leipzig—rather than in France.[4] The significance of the

---

[1] *La filosofia di Giambattista Vico*, pp. 247–8.

[2] R. Flint, *Vico* (London, 1884). Probably the first English discussion of Vico to bring him to the notice of a wide circle of students and teachers was the substantial and very laudatory one in C. E. Vaughan, *Studies in the History of Political Philosophy before and after Rousseau* (Manchester, 1925), i. 207–53.

[3] G. Gentile, *Giambattista Vico* (Florence, 1936); F. Amerio, *Introduzione allo studio di G. B. Vico* (Turin, 1947); T. Berry, *The Historical Theory of Giambattista Vico* (Washington, 1949); A. R. Caponigri, *Time and Idea: The Theory of History in Giambattista Vico* (London, 1953).

[4] A good short demonstration of this, with references to recent German work on the subject, can be found in M. Raeff, 'The Enlightenment in Russia and Russian Thought in the Enlightenment', in J. G. Garrard (ed.), *The Eighteenth Century in Russia* (Oxford, 1973), especially pp. 37–40.

Sturm und Drang writers of the 1770s as forerunners not merely of romanticism but still more of nineteenth-century realism has been grasped.[1] Above all the interest and importance of the Italian contribution to enlightened Europe's stock of ideas has become, in the last few decades, much clearer than ever before. Until recently, at least to the world outside Italy, the Italian Enlightenment did not mean very much more than Cesare Beccaria's *Delle delitti e delle pene*, the little book whose publication in 1764 began the development of modern penology. The existence of other thinkers and writers of interest and even importance—Genovesi, Verri, Galiani, Filangieri, earlier and in a rather different way Giannone—was known; but there was little real knowledge of their works or effort to analyse their relationship to the Enlightenment elsewhere in Europe. So far as the English-speaking world was concerned there was some understanding of Italy's contribution to the development of literary and critical theory in the early eighteenth century, in particular to the concept of the creative imagination which was later to be used in England and Germany as a weapon against classicism and pseudo-classicism.[2] But no comprehensive treatment of the intellectual life of eighteenth-century Italy existed in English.[3] Nor, indeed, does any exist today. Yet the Italian Enlightenment is, for the English reader, slowly emerging from the obscurity and neglect which until recently enshrouded it and taking its rightful place in the intellectual history of Europe. To this development specialized studies by English or American authors have made some contribution.[4] The largest of all, however, has come from Franco Venturi, the leader of Italian historians in this field, in his *Italy and the Enlightenment: Studies in a Cosmopolitan Century* (London, 1972). This collection of essays frequently demonstrates Venturi's admirable tendency to link intellectual developments and controversies with concrete political, economic, or social situations and problems. It provides plentiful

---

[1] Pascal, *The German Sturm und Drang*, p. 313.

[2] J. G. Robertson, *Studies in the Genesis of Romantic Theory in the Eighteenth Century* (Cambridge, 1923), *passim*.

[3] Emiliana P. Noether, *Seeds of Italian Nationalism, 1700–1815* (New York, 1951), though containing useful and interesting material, is a discussion only of a single theme.

[4] An important example is P. Chorley, *Oil, Silk and Enlightenment; Economic Problems in XVIIIth Century Naples* (Naples, 1965), which links the ideas and reform projects of the Neapolitan *illuministi* closely with the economic problems of southern Italy. It deals, however, entirely with the last years of the eighteenth century.

evidence of the very real intellectual activity which characterized much of eighteenth-century Italy.

Even more, however, the historiography of recent years has shown a new width in its view of the Enlightenment by a growing insistence on placing it firmly in its political and social context. It has become increasingly difficult to see it as an autonomous movement of ideas, a wave of intellectual striving and aspiration little connected with the material realities of its own age. Cassirer over forty years ago pushed such an analysis, a treatment of the Enlightenment in terms of pure ideas, to a pitch which has never been surpassed: but the writing of the last decade or more has found this approach to the subject more and more limiting and sterile. The history of ideas, Venturi has proclaimed, 'can be satisfied less and less with the apparent logic of intellectual forms and increasingly seeks to understand what the terms, words, concepts and myths really mean'.[1] A younger writer has attacked 'the old and lazy idealist conception of the Enlightenment as an autonomous spiritual force, which hardly requires explanation'.[2]

This increasing insistence on bringing the Enlightenment down to earth, on seeing it merely as one part of a complex and many-sided historical situation, has had several results. It has meant a movement away from concentration merely on a small number of 'great thinkers' and a growth of interest in 'climates of opinion' and relatively minor figures (in France, Morelly, Mably, La Mettrie, or Linguet, for example). It has meant greater interest than ever before in the mechanisms by which the Enlightenment and its ideas were propagated; in the press, publishing, literary and scientific societies.[3] Most important of all, it has led to much greater efforts than in the past to link closely the movement of ideas, above all in France, with the whole background to those ideas, the political, social, and economic struggles and problems which deeply influenced their scope and form. A considerable effort in this direction was made in the work of Paul Hazard a generation or more ago. The same effort, in more direct and explicit forms, is visible in much modern writing. Professor Hampson, in what is now the most balanced and up-to-date account

---

[1] 'The European Enlightenment' in *Italy and the Enlightenment*, p. 2.

[2] Chorley, *Oil, Silk and Enlightenment*, p. 9.

[3] Perhaps the best illustration of this is the amount of detailed attention paid to the production and publication of the *Encyclopédie* in such works as J. Proust, *Diderot et l'encyclopédie* (Paris, 1962) and J. Lough, *Essays on the 'Encyclopédie' of Diderot and d'Alembert* (London, 1968).

of the Enlightenment in English, is careful to place the intellectual developments with which he is concerned against a very competently drawn background of political and social history.[1] Much of Venturi's work benefits in the same way. A similar linkage, an attempt to use the facts of the external world to explain intellectual attitudes and beliefs, can be seen, often very interestingly, not merely on a wide canvas but also in matters of detail. It has been argued, for example, that enlightened criticism of the debilitating effects of Christianity of the kind which Voltaire and Gibbon in different ways typify, was encouraged by the mere physical state of the city of Rome as it was in the eighteenth century—dirty, decaying, with perhaps a third of the population of the great days of its imperial fame.[2] Another writer has shown convincingly how much of the radicalism of Parisian literary life in the decade or more before the Revolution was not the product merely of intellectual conviction but was rather rooted in a social environment which seemed on every side to deny opportunity to the ambitious young journalist or writer. Envy of privilege, resentment of their own lack of success, bred among such men, as the tide of the Enlightenment ebbed away, a 'gutter Rousseauism' which was a political and above all social rather than an intellectual phenomenon. 'It was from such visceral hatred, not from the refined abstractions of the contented cultural élite, that the extreme Jacobin revolution found its authentic voice.'[3] The most complete, though as yet still insufficiently known, effort to establish on a large scale anywhere in Europe the connection between the movement of ideas and that of political affairs has been provided by an Italian historian. Furio Diaz, in his *Filosofia e politica nel settecento francese* (Turin, 1962) has studied in detail and with remarkable success this relationship in France between the crisis provoked by Machault's new taxation proposals in 1749 and the fall of Turgot in 1776. In almost seven hundred heavily-documented pages the intellectual history of the period is integrated more closely than ever before with the political events and problems of these years. The outcome is the most telling demonstration hitherto achieved of the advantages of writing intellectual history against a political background which is not merely

[1] N. Hampson, *The Enlightenment* (The Pelican History of European Thought, vol. 4, Harmondsworth, 1968), especially chaps. 1 and 5.

[2] R. Pomeau, *L'Europe des lumières: cosmopolitisme et unité européene au dix-huitième siècle* (Paris, 1966), p. 108.

[3] R. Darnton, 'The High Enlightenment and the Low-Life of Literature in pre-revolutionary France', *Past and Present*, No. 51 (May 1971), p. 115.

sketched but made an essential and equal part of the whole story.

It is along this line that fruitful writing about the Enlightenment seems most likely to develop in the future. Voltaire, Rousseau, Montesquieu, even Diderot and Herder, may now have yielded most of their secrets to the questioning of generations of scholars. But in the linking of event and idea, the analysis of the relationship between political practice and intellectual change, there are still large opportunities waiting to be grasped.[1] Study of the Enlightenment, like that of every major aspect of the history of eighteenth-century Europe, has in recent decades widened its scope, become more all-embracing. It has thus become more intellectually satisfying and more in tune with the realities of the subject.

---

[1] For example that offered by a study of the roots of Deism in English republicanism, and of its importance in the transmission of republican aspirations to various parts of continental Europe, as suggested by Venturi, 'The European Enlightenment', pp. 5–8.

CHAPTER III

MONARCHS AND DESPOTS

THE IDEA OF ENLIGHTENED DESPOTISM

ENLIGHTENED DESPOTISM, *despotisme éclairé, aufkgeklärte Despotismus,* is one of the aspects of eighteenth-century Europe to which it is most difficult to give precise meaning.[1] The term itself dates only from the second quarter of the nineteenth century. The nature of the reality it attempts to describe, its dimensions in space and time, perhaps even its very existence, have all been and still are questioned by historians. Hardly any eighteenth-century writer formulated a systematic theory of enlightened despotism. The *L'Ordre naturel et essentiel des sociétés politiques* of the physiocrat Le Mercier de la Rivière (Paris, 1767) is often quoted as the first and indeed almost the only such formulation. But this book, which sees the despot as the guarantor of natural freedom since he declares the laws immanent in the natural order of things, did not command the complete assent of all the political commentators of its own day and would have won even less from those of the earlier decades of the century.[2]

How far was enlightened despotism anything effectively new in the Europe of the middle and later eighteenth century? Clearly many of the policies which the 'enlightened despots' attempted—a reduction in clerical power and privileges; the creation of more efficient and particularly more effectively centralized administrations; the weakening and undermining, to this end, of traditional bodies intermediate between the ruler and his subjects, such as provincial estates; legal codification and the carrying-out of large-scale surveys of landownership; in a few cases even the extension of elementary education to the mass of the ruled—had also been attempted by prede-

---

[1] An anthology of the widely varying definitions provided by different writers can be found in M. Lhéritier, 'Le Despotisme éclairé de Frédéric II à la révolution française', *Bulletin of the International Commission of Historical Sciences,* ix (1937), 193–6.

[2] On the lack of any systematic theory of enlightened despotism see R. Derathé, 'Les Philosophes et le despotisme', in P. Francastel (ed.), *Utopie et institutions au XVIIIe siècle: le pragmatisme des lumières* (Paris–The Hague, 1963).

cessors who did not lay claim to any particular enlightenment and
who did not pretend to be responding to anything more than the
demands of necessity.[1] To distinguish meaningfully between
'enlightened despots' and earlier 'empirical despots' (of whom
Peter I of Russia is an obvious example) may often be difficult. It is
correct, so far as it goes, to describe enlightened despotism as a
combination of largely traditional governmental objectives with
new and liberating ideas; but the exact nature of the different elements
in the compound and the proportions in which they were combined
are less easy to identify. Le Mercier de la Rivière assumed that there
was a natural social and political order which could be easily
discovered and which all enlightened men would support, that a
natural harmony of interests could be achieved without much
difficulty under the rule of an enlightened despot. But these beliefs
and the optimism which underlay them were only in a few cases more
than an agreeable aspiration so far as the rulers of the great states of
Europe were concerned. Sometimes, as in France, vested interests
and traditional routine were too strong to be seriously threatened
by such an ideology.[2] Everywhere at the highest level the realities
remained what they had always been, those of power and of personal
and dynastic ambition. Over most of the Continent society continued
to be dominated by localism and provincialism, by corporate bodies
such as municipalities, guilds, and village communities whose
existence was deeply rooted in history and buttressed by concepts of
natural law.[3] Much modern historical writing as will be seen[4] has
tended to stress the limited practical effect which any 'enlightened'
idea could hope to have on the governments of the great states of
Europe. Sometimes an attempt is made to bridge this gulf between
theory and practice by postulating different forms of enlightened
despotism. That of the great monarchs, of which Frederick II of

[1] M. Lhéritier, 'Le Rôle historique du despotisme éclairé particulièrement au
XVIII⁰ siècle', *Bulletin of the International Commission of Historical Sciences*,
i, Pt. v (1928), 601–12. For the efforts of some historians to push the origins of
enlightened despotism back into the Middle Ages or even farther see the same
author's 'Le Despotisme éclairé', 192–3.

[2] F. Olivier-Martin, 'Des pratiques traditionnelles de la royauté française et le
despotisme éclairé', *Bulletin of the International Committee of Historical Sciences*,
v, Pt. iii (1933), 701–13.

[3] A fact well brought out by D. Gerhard, 'Regionalismus und ständisches
Wesen als ein Grundthema europäischer Geschichte', *Historische Zeitschrift*,
clxxiv (1952), 307–37.

[4] See pp. 152–4, 169–71 below.

Prussia was the prototype and main inspiration, it is argued, was based on the assumption of the primacy of the state and its interests. That of the physiocrats, on the other hand, gave first place to the right of the individual to personal freedom, property, and security. Moreover different rulers took very different views of their position and thought of enlightened rule in very different terms. The regime of Joseph II in the Habsburg territories, essentially humanitarian and liberal, contrasted very sharply, it is claimed, with that of Frederick II in Prussia, rigid and state-centred.[1] But this approach inevitably tends to undermine what unity the phenomenon as a whole possesses. Under such treatment enlightened despotism becomes no more than a group of monarchs and their ministers coping as best they could with widely varying situations and problems and decorating their differing policies with a facade of 'enlightened' jargon. The author of the best modern general work on the subject is driven to conclude that the only feeling which all the enlightened despots shared was 'their common feeling for the state', and that 'no principle really unites them; no practice endows them with unity'.[2]

The effectiveness of enlightened despotism was also subject to clear geographical limits. It has become something of a truism to allege that it could realize its full potentialities only in small states, socially and otherwise more homogeneous than their great neighbours and freer than they from the internal pressures generated by an active and expansionist foreign policy. This belief can claim good contemporary authority, for the Grand Duke Leopold of Tuscany, the most remarkable ruler in Europe during the later eighteenth century, believed that 'in a small state one can do some good'[3] much more easily than in a large one. One important twentieth century writer has even refused to recognize the existence of enlightened despotism, except in the smaller states of the Continent.[4] Yet the greater enlightenment of small political units as against large ones clearly cannot be set up as a general rule in eighteenth-century Europe. Baden or still more the Grand Duchy of Tuscany offer remarkable examples of what might be attempted by an enlightened ruler working within a small and manageable compass. But even in Italy, now increasingly

[1] M. Lhéritier, 'Le Despotisme éclairé', 197 ff., 203–4.
[2] F. Bluche, *Le Despotisme éclairé* (Paris, 1968), p. 353.
[3] A. Wandruszka, 'Pietro Leopoldo e le sue riforme in Toscana', *Archivio storico italiano*, Anno cxviii (1960), 291.
[4] P. von Mitrofanov, *Joseph II; Seine politische und kulturelle Tätigkeit* (Vienna, 1910), p. 83.

seen as an important focus of enlightened speculation and theorizing, there were many small states—Venice, Genoa, Piedmont, the Papal State—in which enlightened reform made little or no headway. Moreover it is upon these states that a good deal of the attention of historians has been concentrated during the last generation; this may be seen as another indication of the tendency of much recent writing to be sceptical of the real significance of enlightened despotism in eighteenth-century Europe.[1]

The phenomenon, then, is irritating in its fluidity. It changes its shape as we grasp it and even threatens to dissolve before our eyes. So far as most rulers were concerned it was perhaps little more than a vaguely defined set of attitudes which could sometimes be used to justify policies which would have been embarked upon in any case. To the *philosophes* it was often merely a weapon for use against clerical power and religious convention, a means of breaking the traditional alliance of throne and altar and making Churches everywhere effectively subject to the state. Certainly it was this anti-clerical animus above all which underlay their extravagant and often sycophantic adulation of Frederick II, Joseph II, and Catherine II.[2]

Yet the scepticism of the preceding paragraphs is only one side, though an important one, of the picture. Rightly or wrongly, most intelligent men during the eighteenth century, and particularly during its second half, believed firmly in the power of an enlightened and energetic ruler to give new form and direction to the political mechanism, and even the society, which he ruled. 'The state', wrote the historian Schlözer at the end of the 1760s, 'is a lifeless mass, to which the monarch first gives life; a machine without motion, which the monarch first sets moving and to whose mechanism he gives reality . . . the monarch lives for the state and the state lives through him.'[3] This was an extreme statement, but it reflected a widely held attitude. Several factors underlay this acceptance of and even

---

[1] R. Villari, 'Il riformismo e l'evoluzione delle campagne italiane nel settecento, attraverso gli studi recenti', *Studi storici*, Anno v (1964), No. 4, 628–9. Such works as M. Berengo, *La società veneta alla fine del settecento* (Florence, 1956) and G. Quazza, *Le riforme in Piemonte nella prima metà del settecento* (Modena, 1957), bring out well the unimportance in practical terms of the Enlightenment in much of Italy and the severely limited nature of much of what reform was introduced there.

[2] R. Derathé, 'Les Philosophes et le despotisme', pp. 69–70.

[3] A. L. von Schlözer, *Neuverändertes Russland oder Leben Catharina der Zweyten, Kaiserinn von Russland* (3rd edn., Riga–Mittau–Leipzig, 1771–2), i, *Vorrede*.

enthusiasm for rule by powerful but enlightened monarchs. States in which monarchical power was diluted, either by the existence of deeply-rooted corporate institutions, as in France, or by the strength of the nobility, as in Sweden and, above all, in the Polish Republic, seemed to lag, sometimes dangerously, behind their neighbours in political and military efficiency. A strong, even despotic, monarchy under an enlightened ruler seemed to make possible rapid progress of a kind not to be achieved in states in which power was more widely distributed; and this assumption was strengthened by a prevalent view of history as the story of the deeds and achievements of a small number of great men. Moreover, there was a clear realization that the 'despotism' of Frederick II or Catherine II was something very different from the oppressive and completely arbitrary oriental despotisms condemned by political writers from Montesquieu downwards; a sincere belief that the best form of government was one subject to fixed laws did not by any means rule out the possibility of these laws being the work of an enlightened ruler. Ideas of this kind never won unanimous acceptance. In France particularly they were rejected by many thinkers, notably by Diderot, while Rousseau, in spite of all the authoritarian possibilities inherent in his thinking, opposed strongly the despotism of any individual. Nevertheless it is clear that however cloudy and shapeless an entity enlightened despotism may seem to the present-day historian the hopes and assumptions which underlay it had real meaning for most of the educated and politically conscious inhabitants of continental Europe in the second half of the eighteenth century.[1]

None the less, the historians of the nineteenth century were very slow to recognize enlightened despotism as a distinct historical phenomenon. The term plays no significant part in the writing of the century on Frederick II, Catherine II, or Joseph II. Nor does it figure at all in the most widely-read works of the period on the states and rulers of Mediterranean Europe.[2] The first use of it seems to

---

[1] The best short account of the approach of the philosophes to the question of despotic rule and its possibilities is to be found in A. Lortholary, *Le Mirage russe en France au XVIIIe siècle* (Paris, 1951), pp. 135–50.

[2] Thus William Coxe was able to provide, in his *Memoirs of the Kings of Spain of the House of Bourbon* (2nd edn., London 1815), v. 214–18, a run-of-the-mill summary of the character and achievements of Charles III which betrays no realization whatever that anything that could be called enlightened despotism had been functioning in Spain during his reign. At the other end of the century Dedevizes du Désert, in his *L'Espagne de l'ancien régime* (Paris, 1897–1904) showed an equal unconsciousness of the phenomenon. Again Pietro Colletta, in

occur in the 'Umrisse zur Naturlehre der drei Staatsformen' of the historian and economist Wilhelm Roscher, which was published in the *Allgemeine Zeitschrift für Geschichte* in 1847. In this article he envisages the great reforming monarchies of the later eighteenth century as the successors of the 'confessional absolutism' of Philip II and Ferdinand II in the sixteenth and early seventeenth centuries and of the later 'court absolutism' typified above all by Louis XIV. Whatever may be thought of this general scheme it seems clear that Roscher was the writer who first posited sharply the existence of a new, distinct, and identifiable type of monarchy throughout much of Europe during the generation or more before the French Revolution. From him there also stems a tendency, still active in much historical writing, to regard regimes of this kind as a distinctively German phenomenon or at least as one with mainly German roots. The overriding importance of Frederick II as the first true enlightened despot, the ruler who presented Europe with a model of progessive monarchy which his fellow-monarchs attempted to imitate, is central to this interpretation. In 1889 this view of the primacy of Frederick as the archetype, indeed the creator, of enlightened despotism, was given influential expression in an article by his biographer Reinhold Koser which was one of the first scholarly examinations of the question. More than four decades later the same view, buttressed by the research done in the interval, could be repeated in the same periodical in an identically-titled article by an equally great scholar, Fritz Hartung.[1] Today it is still deeply influential. Bluche, for example, asserts flatly that 'without Frederick II there is no enlightened despotism' and that 'enlightened monarchy is a creation of Frederick II'.[2]

Only in the early part of this century did the idea of enlightened despotism as a useful historical category begin to make headway in writing published outside Germany. With this development, however, came at once a challenge to the view of this form of monarchy as merely or essentially an export from Prussia. It now began to be seen

---

his *History of the Kingdom of Naples, 1734–1852* (Edinburgh, 1858), the most widely-read book of the century on the recent history of southern Italy, though he described the reforms of Tanucci and his successors made no effort to link them with developments elsewhere in Europe or to show them as more than a purely local phenomenon.

[1] R. Koser, 'Die Epochen der absoluten Monarchie in der neueren Geschichte', *Historische Zeitschrift*, lxi (1889), 246–87; F. Hartung, 'Die Epochen der absoluten Monarchie in der neueren Geschichte', *Historische Zeitschrift*, cxlv (1932), 46–52.

[2] *Despotisme éclairé*, pp. 100, 337.

as the practical reflection of the ideas of the Enlightenment, ideas which had their centre unmistakably in France.[1] The chequered relationship between Frederick II and Voltaire, the ostentatious admiration of Catherine II for the leading French thinkers of the 1760s and 1770s, could be used to support this point of view. It was, nevertheless, one very much open to criticism, for in many parts of Europe there were indigenous roots from which enlightened despotism could develop. In these areas it was tempting to see it as essentially a native growth independent of French influences. Frederick II, it could be argued, drew his ideas from Locke, Leibniz, and Christian Wolff, and from the traditions of government built up in Brandenburg by his predecessors, far more than from any writer of the Enlightenment in France. In southern Europe neither Charles VII in Naples nor the Grand Duke Leopold in Tuscany had been deeply influenced by French ideas. Very often memories of Louis XIV and his achievements had been a more powerful influence emanating from France than anything the *philosophes* could provide.[2] In Italy particularly there has been in much recent writing a tendency to stress the indigenous roots of the Enlightenment there and consequently those of the efforts at administrative, legal, and governmental reform which marked at least a few states during the later eighteenth century. The negative view taken by Benedetto Croce, the most influential Italian historian and philosopher of the early twentieth century, of the intellectual achievements of the Italian Enlightenment (he criticized particularly the unhistorical character of its thinking) for long influenced the attitude of other Italian scholars. It strengthened the tendency to regard the Enlightenment, and thus the efforts at practical reform associated with it, as essentially imports from France, excessively rationalistic and abstract.[3] This view, however, has always been challenged by a competing one which sees it in the Italian states as a movement native to the soil and as putting forward ideas in the main indigenous to the peninsula;[4] and the second tendency has on the whole been the

[1] Lhéritier, 'Le Despotisme éclairé', p. 190.

[2] For a recent discussion of these points see Bluche, *Despotisme éclairé*, pp. 26, 210, 341–5.

[3] e.g. F. Lemmi, *Le origini del risorgimento* (Milan, 1924), p. 7.

[4] e.g. E. Rota, *Le origini del risorgimento* (Milan, 1938), i. 84–98, 160–200, 296–300. See also the comments on recent Italian writing, and particularly on the importance of the work of Venturi in this connection, in P. Villani, *Feudalità, riforme, capitalismo agrario: Panorama di storia sociale italiana tra sette e otto-cento* (Bari, 1968), p. 27.

predominant one in the writing of the last generation, though its sway is by no means unchallenged.[1]

Imitation of the achievements of Frederick II, admiration of the ideas of the Enlightenment in France, the natural working of forces indigenous to the states concerned: to this range of explanations of enlightened despotism can be added another—the Marxist or Marxist-influenced one. This has never taken a rigidly stereotyped form (partly no doubt because neither Marx nor Engels ever showed much interest in the forms of government of eighteenth-century Europe). It rests, however, on the assertion that the enlightened despots fostered, for the most part unconsciously, the interests of a middle class aware of its growing economic strength but still in a politically subordinate position from which it saw no hope of freeing itself except with the help of an enlightened ruler. No such ruler, it is agreed, had any idea of establishing an equality of rights between middle class and nobility: every monarch depended much too heavily on the support of the latter for this to be a practical possibility. But in east and central Europe in particular the enlightened despots formed a kind of unspoken alliance with the bourgeoisie, whose growth they favoured. Enlightened despotism thus formed 'an intermediate stage between arbitrary tyranny and bourgeois monarchy'.[2] This is not a view which historians in the English-speaking countries have in general found convincing. There seems to be little clear correlation in any of the European states between the existence of a form of government which can be even plausibly described as 'enlightened despotism' and a strengthening of the middle class. In Piedmont, for example, the administrative machine became dominated, even in its higher ranks, by middle-class office-holders during the first half of the eighteenth century: the nobility was pushed in this respect almost completely into the background.[3] Yet the country's rulers and system of government can hardly be seen as 'enlightened' during this period. On the other hand in Naples, where 'enlightened'

---

[1] See for example F. Diaz, *Francesco Maria Gianni: della burocrazia alla politica sotto Pietro Leopoldo di Toscana* (Milan–Naples, 1956), Introduction, for an illustration of the strength of French influences in the Enlightenment in Tuscany.

[2] This very brief summary is based mainly on G. Lefebvre, 'Le Despotisme éclairé', *Annales historiques de la révolution française*, n.s., xxi (1949), 97–115, which is a short and accessible statement of the Marxist point of view by a great historian.

[3] G. Quazza, *Le riforme in Piemonte nella prima metà del settecento* (Modena, 1957), i, Chaps. II and III.

ideas at every level were far stronger, the most detailed study hitherto undertaken suggests that there was a real decline in the position of the middle class as a result of the failure of the revolt of 1647 against Spanish rule, that this decline had become clear by the end of the seventeenth century and that it did not even begin to be reversed until the last decades of the eighteenth.[1] In Spain, however, it seems that from the 1760s onwards, under the somewhat half-hearted enlightened despotism of Charles III and his ministers, both the middle class and the great landed nobility were enjoying a period of economic well-being and that friction between them was as a result relatively slight.[2] In Russia, on the other hand, according to many Soviet historians, even the foundations of genuine capitalism, without which a substantial modern middle class is hardly conceivable, were merely beginning to be laid in the 1760s.[3] The picture is thus a very cloudy and imprecise one. Underlying any specific inconsistencies and discrepancies, moreover, is the more profound difficulty of discussing any aspect of government in class terms in an age when classes of a modern kind scarcely existed in most European states and the 'middle class' in particular was a very heterogeneous entity.[4]

There are two further characteristics of writing on enlightened despotism during the last generation which deserve mention. The first is a marked tendency to concentrate on the practical aspects of the phenomenon, to study the workings of this system of government (if this is not too precise and clear-cut a phrase) and to pay less attention to the theories which may have underlain it. In Italy this turning away from the ideas and intellectual background of enlightened rule and towards the study of its practical achievements has been particularly marked. In the early years of this century Croce

---

[1] See the second part of L. De Rosa, *Studi sugli arrendamenti del regno di Napoli. Aspetti della distribuzione della richezza mobiliare nel mezzogiorno continentale (1649–1806)* (Naples, 1958).

[2] R. Herr, *The Eighteenth-century Revolution in Spain* (Princeton, 1958), p. 232.

[3] H.-J. Torke, 'Die Entwicklung des Absolutismus-Problems in der sowjetischen Historiographie seit 1917', *Jahrbücher für Geschichte Osteuropas*, Neue Folge, Band 21 (1973), 501.

[4] For a discussion of these difficulties in the context of enlightened despotism see F. Diaz, 'Recenti interpretazioni della storia della Toscana nell'età di Pietro Leopoldo' *Rivista storica italiana*, lxxxii (1970), 387–99. A more general view of the pitfalls of writing the history of any period before the nineteenth century in class terms can be found in *Problèmes de stratification sociale. Actes du colloque international (1966) publiés par Roland Mousnier* (Paris, 1968).

helped to focus the attention of scholars concerned with eighteenth-century Italy in the main on the intellectual development of the peninsula. This he achieved in particular through the great influence and prestige of his *Istoria del regno di Napoli* (Bari, 1925) a book which, as a recent critic has complained, 'reduces the whole history of the Mezzogiorno to that of the intellectual class'.[1] The highly intellectualized approach to the subject which he so remarkably typified has, however, been more and more discarded in recent years in favour of that represented most notably by Franco Venturi. This involves the study of the impulse towards reform in eighteenth-century Italy not in terms of political ideas or abstractions of any kind but in those of concrete problems, needs, and achievements. This has meant on the one hand a more intensive study than any hitherto attempted of the efforts towards legal reform and economic development which were an important aspect of the Italian Enlightenment. On the other it has led to much detailed work on the lives and activities of individual reformers.[2] Nor has this interest in practicalities been confined to Italy. There have been calls by historians elsewhere in Europe for more attention to the working in practice of theories of enlightened government, for more study of the enormous difficulties posed for would-be enlightened rulers by their frequent lack of efficient subordinates and by the inadequate communications which afflicted every large eighteenth-century state. The relationship between theory and practice in such fields as charity and poor relief, education and justice, has been seen as urgently in need of more searching investigation.[3] This attitude has borne fruit in the production of a number of important books which attempt, more comprehensively and in greater detail than ever before, to see both the enlightened despotism of the later eighteenth century and the societies in which

---

[1] P. Villani, 'Risultate della recente storiografia e problemi della storia del regno di Napoli', in the same author's *Mezzogiorno tra riforme e rivoluzione* (Bari, 1962), p. 48. Croce does in fact speak of the educated upper middle class in Naples as embodying the critical faculty and the will to progress and reform, and also as in a sense representative of the people of the kingdom as a whole, a claim with which no present-day historian would agree (*Istoria del regno di Napoli* (3rd edn., Bari, 1944) pp. 176 ff.).

[2] The series *Illuministi italiani*, edited by Venturi, gives a more dramatic picture of the progress of practical reform and the obstacles it had to face than any to be found in older works. See in particular vol. iii, *Riformatori lombardi, piemontesi i toscani* (Milan–Naples, 1958) and vol. v, *Riformatori napoletani* (Milan–Naples, 1962).

[3] For a good statement of this point of view, see G. Livet, 'Introduction à une sociologie des lumières', in P. Francastel (ed.), *Utopie et institutions*, pp. 267–8.

it had to work in terms of realities rather than of theories or ideas, in terms of the detailed and the concrete rather than the general.[1]

This tendency has had as its main result a playing-down of what elements of novelty were present in enlightened despotism. All that was new, it can well be argued, about the enlightened monarchs of the later eighteenth century was the ideas which may have inspired them. Their practice was in essentials in the tradition of their predecessors. Many of the policies of Catherine II (her efforts at legal codification are a good example) were a direct continuation of those of earlier rulers of Russia; and she was seen by some of her contemporaries as inheriting the mantle of Peter I.[2] In the same way the administrative reforms of Joseph II were built on those, in many ways more far-reaching, of his mother; and 'Josephinism' in religious matters and church–state relations can be traced back to the very early years of the eighteenth century.[3] Frederick II was dominated by, indeed the prisoner of, assumptions and necessities deeply-rooted in the history of Brandenburg–Prussia. In many fundamental respects he was able merely to develop and elaborate policies laid down by his father. The more enlightened despotism is looked at simply in terms of its workings, the more it tends, at least in the greater states, to become dissociated from the Enlightenment and to seem the outcome either of history or of the practical necessities facing the rulers concerned.

One recent writer has drawn a very clear distinction between the Enlightenment, abstract, cosmopolitan, believing in the essential goodness of man and aiming at the general happiness, and the monarchies of the later eighteenth century, dominated by concrete ambitions, often nationalist, frequently imbued with a deep contempt for the ordinary man and concerned above all with the defence and extension of their own power.[4] This is an oversimplification. But more and more in recent years historians have been tempted to stress the continuities which link the enlightened despots with their predecessors rather than to see these rulers as creating any genuinely new form of rule. 'Taken as a whole,' wrote Lefebvre a generation ago, 'the practice of government in the eighteenth century did not

---

[1] An outstanding example in the context of Spain, a state hitherto in many respects rather badly served by historians of the eighteenth century, is J. Sarrailh, *L'Espagne éclairée de la seconde moitié du XVIIIᵉ siècle* (Paris, 1954).

[2] See below, p. 160.

[3] See below, pp. 184–7

[4] Bluche, *Despotisme éclairé*, pp. 331–3.

differ in principle from what it had been earlier.'[1] A few years later Hartung and Mousnier, with the advantage of an even deeper knowledge of the subject, came to the same conclusion: 'In sum, the "enlightened despots" did what other absolute rulers had done before them.'[2] This is a verdict with which the majority of historians would today find themselves in general agreement. It is also one increasingly supported by the study of individual states and rulers. The 'enlightened despotism' of Charles III of Spain, for example, can now be seen not as something new, the outcome of ideas imported from France, but rather as part of a long-standing Spanish tradition which hoped radically to reform the country by the application of a few simple and far-reaching ideas. This, it now seems clear, was an outlook already clearly expressed in many of the proposals for change made in the seventeenth and early eighteenth centuries. Even the regalism, the desire to assert royal authority over the church, shown by some of Charles's ministers such as Campomanes, had deep roots in Spanish history. The policies of Charles and his regime were thus in practice cautious and largely traditional.[3] In Naples, again, many of the policies of Charles III after his establishment in power there in 1734 had much more than is usually realized in common with those of the unenlightened Austrian regime which had preceded him.[4] The long anti-papal tradition behind many of the policies of Joseph II has been clarified in a series of works by Eduard Winter,[5] while the judgement of Hartung and Mousnier that the internal policies of Frederick II were 'conservative to the point of immobilism'[6] is merely a somewhat extreme statement of a point of view which almost all historians would today accept.

Looked at under the magnifying-glass and considered in terms of deeds rather than words, enlightened despotism thus shrinks to

---

[1] 'Le Despotisme éclairé', p. 103.

[2] F. Hartung and R. Mousnier, 'Quelques problèmes concernant la monarchie absolue', *X Congresso internazionale di scienze storiche; Relazioni* (Florence, 1955), iv, 17.

[3] M. Desfourneaux, 'Tradition et lumières dans le "despotismo illustrado" ', in P. Francastel (ed.), *Utopie et institutions*, pp. 229–45. The tendency of recent Spanish writing to stress the more constructive aspects of Spain's history in the seventeenth century and to reject the view of it as a period of complete decadence is well brought out in L. Sanchez Agesta, *El pensamiento politico del despotismo illustrado* (Madrid, 1953).

[4] Villani, 'Risultate della recente storiografia', pp. 9–12.

[5] See below, pp. 187.

[6] 'Quelques problèmes', p. 17.

little more than a set of theories and aspirations which were used, at least in the greater states of Europe, to give an intellectual veneer to policies which were seldom genuinely new and frequently selfish. The reality of much political life in eighteenth-century Europe was that of the monarchs, their personalities, ambitions, and limitations. It is these individuals, rather than so vague and unsatisfactory a subject as enlightened despotism in general, who have since their own day attracted most of the attention of historians; and it is to the most important of them that we must now turn.

PRUSSIA: FREDERICK THE GREAT

No ruler of the eighteenth century, perhaps none in modern times, has been more written about than Frederick II of Prussia. A bibliography published on the centenary of his death could already list about 2,700 books and articles relating to him;[1] today the number would be enormously greater. Yet this flood of writing has probably impeded as much as it has helped the achievement of a completely satisfactory analysis of his work and significance. Nor have his own character and personality done anything to ease the task of the historian. More than most human beings he was made up of contradictions, torn between a genuine taste for an intellectual and artistic life largely divorced from the day-to-day world and a deep-seated drive towards action and practical achievement, between a rooted and cynical realism on the one hand and an urge to gamble and take great risks for high stakes on the other.[2] The sheer scope of his interests and activities, wider even than those of his contemporary Catherine II of Russia, is another obstacle in the way of the historian who tries to produce a comprehensive account of his achievements. Above all, he illustrates the truth of the claim that every age writes or rewrites history in its own image and from the standpoint of its own assumptions and preoccupations. There is no great historical figure of whom the judgement of posterity has

---

[1] M. Baumgart, *Die Literatur des In-und Auslandes über Friedrich den Grossen* (Berlin, 1886). The bibliographical literature on Frederick is extensive. Two useful though very summary discussions in English of the more important books about him are: V. Valentin, 'Some Interpretations of Frederick the Great', *History*, xix (1934-5), 115-23; and G. P. Gooch, *Frederick the Great* (London, 1947), chap. xv.

[2] These contradictions are briefly discussed in P. Paret (ed.), *Frederick the Great: A Profile* (London, 1972), Introduction, pp. xiii-xvii.

fluctuated more violently. Passionately admired by the majority of his contemporaries as the supreme model of enlightened despotism[1] he was within hardly more than a decade of his death coming under the lash of the German romantics. The object of adulation in Bismarckian and Wilhelmine Germany, his historical reputation was shaken by the defeat of 1918 and seriously undermined by the catastrophe of 1945. In Germany and to a lesser extent in other countries, each age has seen the Frederick whom it wished to see and has stressed in him the elements (inventing them when they did not exist) which served its own purposes and suited its own prejudices. But however admiring or hostile the estimate of him, his immense importance in the history of Germany and indeed of Europe has scarcely been questioned. Until recently it would have been difficult to find many historians, in Germany or elsewhere, who disagreed with the assertion of one of his modern admirers that 'All Prussians have in fact been shaped by this ruler'.[2]

The judgements passed on him during his lifetime were usually favourable; those of the numerous books called forth by his death were even more markedly so.[3] The attractions of success can here be seen clearly at work. In all the great crises of his reign, it was generally agreed, his behaviour had been justifiable, even praiseworthy. The Hohenzollerns had had in 1740 good claims to most of Silesia, claims which it was an obligation of honour to enforce. In 1756 Frederick, threatened by an overwhelming hostile coalition, had had no alternative but to take the initiative and attempt to forestall his enemies. The first partition of Poland had been the work of Catherine II in which Prussia had been forced to cooperate.[4] Criticism was not completely lacking. The military character

---

[1] See the comments of R. Koser, 'Staat und Gesellschaft zur Hoheit der Absolutismus', in *Staat und Gesellschaft der neueren Zeit* (Berlin–Leipzig, 1908), p. 240; and the more recent and perhaps too sweeping assertions of F. Bluche, *Le Despotisme éclairé*, pp. 28, 100.

[2] H. J. Schoeps, *Die Ehre Preussens* (Stuttgart, 1951), p. 14.

[3] For a detailed study of contemporary French comment on Frederick see S. Skalweit, *Frankreich und Friedrich der Grosse: Der Aufstieg Preussens in der öffentlichen Meinung des 'ancien régime'* (Bonn, 1952). English estimates of him during the first half of his reign are well discussed in M. Schlenke, *England und das friderizianische Preussen, 1740–1763* (Freiburg–Munich, 1963). Emmy Allard, *Friedrich der Grosse in der Literatur Frankreichs, mit einem Ausblick auf Italien und Spanien* (Halle, 1913) is a study of contemporary and near-contemporary literary judgements.

[4] (J. C. T. de Laveaux), *The Life of Frederick the Second, King of Prussia* (London, 1789), i. 38, 118–23, ii. 133, 275: J. Gillies, *A View of the Reign of*

of the Prussian government and the heavy taxation it involved; Frederick's desire for personal power and occasional arbitrariness; the mediocrity of many of his ministers—all these were mentioned by at least some commentators.[1] But they seemed minor defects when set against his achievements; in particular his passion for work and devotion to duty could be contrasted very favourably with the dilatory self-indulgence of his successor, Frederick William II, a point made repeatedly by the most penetrating of late-eighteenth-century observers, the Comte de Mirabeau.[2] 'The mass of good he did', proclaimed an admirer, 'outweighed the bad'; and 'nature, to form Frederick the Great, used the substance of four illustrious men'.[3] Another claimed that 'he was, during half his reign, the god of war; we shall see him, during the other half, as the god of peace'.[4] Nor was it only among the educated and well-informed that this fulsome praise was lavished on the newly-dead king. On an intellectually lower and more popular level the collections of anecdotes about him, of which several appeared, reflect the same largely uncritical approval.[5]

By the end of the century it was clear that this attitude was changing, at least in Germany. In this process the hostility to Frederick of his successor, the desire of Frederick William II and his ministers to play down the achievements and reputation of a man whose free-thinking in religious matters they disliked and whose spectacular achievements they could not hope to emulate, were of little importance.[6] Far more significant was the change in the whole temper

---

Frederick II of Prussia (London, 1789), pp. 65–7, 217: Comte P. H. de Grimoard, *Tableau historique et militaire de la vie et du règne de Frédéric le Grand, roi de Prusse* (London, 1788), pp. 16, 51–2.

[1] e.g. Laveaux, *Life of Frederick the Second*, ii. 255–60.

[2] *Histoire secrète de la cour de Berlin* (Alençon, 1789), i. 215–17, 256 ff., ii. 80 ff., 161. The moral example which Frederick's tireless industry offered to contemporaries is brought out in the comments quoted in R. Schwarz, *Friedrich der Grosse im Spiegel des literarischen Deutschlands von der Aufklärung bis zur Romantik* (Leipzig, 1934), pp. 146–7.

[3] Grimoard, *Tableau historique et militaire*, pp. 30–41.

[4] Comte J. A. L. de Guibert, *Éloge du roi de Prusse* (London, 1787) p. 240.

[5] e.g. *Traits caractéristiques et anecdotes de la vie de Frédéric II* (Strasbourg, 1788); *Frédériciana, ou recueil d'anecdotes, bons mots et traits piquans de Frédéric II* (Paris, 1801).

[6] Except in so far as they may explain the shamefully inadequate edition of Frederick's *Œuvres* which appeared at Berlin in 1788 and which was not superseded for more than two generations. See Schwarz, *Friedrich der Grosse*, pp. 149 ff.

of German intellectual life which was now being produced by the triumph of romanticism. That triumph meant that by the early years of the nineteenth century a galaxy of the most influential German writers—Novalis, Schelling, Arndt, Wackenroder, Adam Müller— were with differing degrees of intensity hostile to Frederick. There were several ways in which he was vulnerable to their criticism. Romantics of a predominantly religious or even mystical cast of mind —Novalis is an obvious example—could attack him as the representative of the eighteenth-century Enlightenment, cold, rational, mechanical, ultimately life-denying. Romanticism of a German nationalist kind, on the other hand, could see in him the ruler who, indifferent or even hostile to national needs and feelings, had proclaimed the superiority of French over German intellectual and artistic life and openly despised the German language. Moreover, had he not fatally weakened the old reich in the face of foreign enemies by his creation of a Prusso–Austrian dualism and a lasting hostility between Hohenzollerns and Habsburgs? In his *Europa und Germanien* (1803) and the first part of his *Geist der Zeit* (written in 1805–6) Arndt put forward these charges with real bitterness. 'Foreign to the spirit of this monarchy', he wrote of the Frederician regime, 'was all that is called German'; and 'there is nothing more laughable than to attribute to Frederick patriotic German ideas', since his dynasty, not the nation, was the object of his solicitude.[1] The conservative and Catholic romantic, Adam Müller, found yet another line of attack. Frederick, he argued in his *Über König Friedrich II und die Natur, Würde und Bestimmung der preussischen Monarchie* (Berlin, 1810), had strengthened the pernicious idea that what mattered in history was great men, whereas it was really made by states and nations, entities which transcended mere individuals and whose progress could not be stultified by the action of individuals.[2]

These were the judgements of poets, mystics, or at best political philosophers, not of historians; and the historical writing of the period was kinder to Frederick than they would suggest. The great school of historians associated with the university of Göttingen—Spittler, Heeren, Pütter—were favourable to him. The Swiss historian Johannes

---

[1] Schwarz, *Friedrich der Grosse*, pp. 175–6. Chap. v of this book, though concerned essentially with literature, is the best discussion of German writing in general about Frederick during the period 1786–1815.

[2] W. Bussmann, 'Friedrich der Grosse im Wandel des europäischen Urteils', in *Deutschland und Europa* (Düsseldorf, 1951), pp. 377–8.

von Müller, the newly-appointed historiographer to the Prussian monarchy, defended his memory in a famous lecture to the Academy of Sciences in Berlin in January 1805, arguing that he had given Prussia an inner strength which allowed it to face the future with confidence and the Prussians a kind of national or at least state consciousness hitherto unknown to them.[1] But the effect of all this was limited. None of the Göttingen historians wrote about Frederick in a serious or large-scale way. The collapse of 1806 was to show that von Müller's praise was at least exaggerated and apparently to justify accusations that the Prussian state which Frederick left his successors was a mere mechanism without real life-force or power of growth and development.[2]

The generation which followed the end of the Napoleonic Wars saw a distinct rise in the king's standing in the eyes of historians and of the general public. Interest in him grew. The first proper historical works on him appeared. Documents relating to his reign began to be published, though as yet on nothing approaching the scale which was to mark the last decades of the nineteenth century. Increasingly, writing on him became dominated by historians or writers with some pretensions to be such; no longer was it the work of pamphleteers or poets. For the first time Frederick's relations with his father, a subject which has continued to preoccupy historians to the present day, began to be seriously investigated.[3] No really profound study of the effect on him of this terrible childhood was attempted; perhaps in the nature of the case no completely satisfactory one can ever be written. But there were now signs of a realization of the crucial importance to him of these early years, an importance which eighteenth-century writers had ignored or glossed over. There was also for the first time a significant output of detailed, comprehensive studies of Frederick and his reign. Some of these were no more than conventional narrative accounts, based on a limited range of printed materials

---

[1] Schwarz, *Friedrich der Grosse*, pp. 151, 168.

[2] On the effects of the defeat of 1806 on judgements of Frederick see R. Koser, *Geschichte Friedrichs des Grossen* (6th and 7th edns., Stuttgart–Berlin, 1921–5), iii. 552.

[3] The best example is the first part (pp. 1–228) of F. Förster, *Friedrichs des Grossen Jugendjahre, Bildung und Geist* (Berlin, 1823), which provides a detailed account of the relations between Frederick William I and his son down to 1739. A small collection of documents relating to the childhood and education of Frederick was published in F. Cramer, *Zur Geschichte Friedrich Wilhelm I und Friedrich II, Könige von Preussen* (2nd edn., Leipzig, 1833), chap. i.

and often heavily biased towards foreign policy and military affairs.[1] Others however were works of real scholarship, efforts of a kind hitherto unknown to treat the subject, or large parts of it, on the basis of unpublished materials and to make new contributions to knowledge. Thus, for example, J. D. E. Preuss, the Prussian official historiographer, produced, in the four volumes of his *Friedrich der Grosse. Ein Lebensgeschichte*, (Berlin, 1832–4), a book which was to provide historians for several decades to come with a secure foundation for further research. Unlike almost all his predecessors he gave something like adequate attention to the internal administration and economic life of Prussia under Frederick,[2] and thus did a good deal to undermine the concentration on war and diplomacy which had unbalanced so much previous writing on the subject. Also by printing a fairly substantial collection of documents at the end of each of his volumes he did more than any of his predecessors to provide historians with new materials on which to base their judgements of the king. A year or two later unprinted materials of a different kind were made available on a hitherto unprecedented scale when Friedrich von Raumer published a study of the international relations of Frederick's reign which was in essentials merely a collection of quotations from the British diplomatic correspondence of the period.[3] A decade later Leopold von Ranke, the greatest historian then alive, published his *Neun Bücher Preussische Geschichte* (Berlin, 1847).[4] This suffers in modern eyes, as an account of Frederick's reign, from the over-concentration on diplomacy which Ranke shared with almost all his predecessors and lesser contemporaries. It was none the less the most balanced and complete estimate as yet made of Frederick's international significance and was based on what was, by the standards of the age, a wide range of materials (including a good deal drawn from the French Foreign Ministry archives).

The judgements passed on Frederick by these writers, both the

---

[1] e.g. C. Paganel, *Histoire de Frédéric-le-Grand* (2 vols, Paris, 1830); T. West, *Friedrich der Grosse* (Berlin, 1839); *Frederick the Great and his Times: edited, with an Introduction, by Thomas Campbell, Esq.* (4 vols., London, 1842–3).

[2] e.g. in the substantial sections on 'König Friedrich als Landesvater und als Mensch' in vol. i, and on 'Friedrich der Grosse nach den siebenjährigen Kriege als Landesvater' in vol. iii.

[3] *Contributions to Modern History from the British Museum and the State Paper Office by Frederick von Raumer. Frederick II and his Times* (London, 1837).

[4] My references are to the English translation, *Memoirs of the House of Brandenburg and History of Prussia during the Seventeenth and Eighteenth Centuries* (3 vols., London, 1849).

purely derivative and the genuine searchers-out of new information, were generally, sometimes overwhelmingly, favourable. When dealing with the events crucial for any assessment of the morality or even the common sense of his policies—the invasion of Silesia in 1740, the attack on Saxony in 1756, the seizure of West Prussia in 1772—they usually returned verdicts in his favour. Austrian jealousy of the rising power of Prussia, and her alleged deception over Prussian claims to the west German duchies of Jülich and Berg, were used to justify Frederick's action in December 1740. In any case, the conquest of Silesia was justified, it could be argued, since 'Prussia needed it, less for aggrandisement than for survival'.[1] In 1756 Frederick had wished for peace but found himself confronted by a ring of enemies who dreamt only of war. The existence of Prussia was thus at stake. 'Mere waiting would have brought Russian armies into Prussia, the French to Westphalia, and the king would not have gained, by neglecting the right moment, the smallest compassion, or any assistance in Europe.'[2] The first partition of Poland was more difficult to justify; and more than one writer referred to it as the only real blot on Frederick's reign.[3] Yet even here there was a consistent failure or refusal to admit that Frederick had desired a dismemberment of Poland. The fact that he had advocated a forcible seizure of West Prussia as early as 1731 was still unknown; and there was a notable eagerness to present Catherine II as the real architect of the 1772 partition and the king of Prussia as merely her more or less reluctant accomplice, driven by the logic of events and the need to defend the interests of his kingdom into policies which he had not foreseen or desired.[4]

Historians so able as Raumer and above all Ranke did not succumb to the temptation to admire Frederick uncritically. The former had doubts about the justification of the Prussian action in 1740 and admitted that the Austrian court had been 'justly indignant' when faced with Frederick's attack. Ranke, though defending the king's abandonment of his French ally in 1742, when he made peace with the Austrians at Breslau, admitted that on this issue 'the judgement of the present time will be by no means favourable to Frederick's

[1] Preuss, *Friedrich der Grosse*, i. 159–62; Ranke, *Memoirs of the House of Brandenburg*, ii. 92, 111–12; Paganel, *Histoire de Frédéric-le-Grand*, i. 211–16, 321.

[2] Raumer, *Contributions to Modern History*, pp. 226, 259, 299–300. cf. Preuss, ii. 3–6; Paganel, ii. 16, 29.

[3] Paganel, ii. 233, 437; *Frederick the Great and his Times*, i, Preface, x–xi.

[4] Preuss, iv. 43–4; Paganel, ii. 225–9.

conduct'.[1] But doubts and hesitations of this kind were submerged by the thought that the rise of Prussia (which almost all of these writers saw as overwhelmingly the work of Frederick alone) had been an essential benefit to Europe, a great step in the march of progress. 'A retrospective review of the last hundred years', wrote Raumer, 'proves to us that Providence had greater ends to accomplish, through Frederick, and the exaltation of Prussia, than could at that time be conceived or presaged.'[2] These ends, to Protestant north-German or British historians, seemed clear. They were the checking of French power and, even more important, of Catholic influence in Europe. If Prussia and Frederick had not existed in 1740, alleged Ranke, France would have achieved permanent hegemony over the Continent. To Campbell, Frederick (and for that matter also his father) stood for religious toleration and the rights of Protestantism throughout Europe. Had Prussia been crushed by her enemies 'Catholicism would have overridden the continent and the Reformed religion would have had to fight another Thirty Years War.'[3] A very genuine fear and suspicion of Catholicism had contributed to Frederick's popularity throughout Protestant Europe in his own lifetime. To it was now being added a quasi-racial sense of northern superiority to southern and Mediterranean Europe.[4] Together these two factors were for the next century to provide support for favourable assessments of the significance of Prussia and of her most obviously successful ruler.

It is true that the 1840s saw the appearance of the most forthright attack on Frederick hitherto made by any historian, with the publication in the *Edinburgh Review*, in April 1842, of Macaulay's famous essay on him. This struck, on many issues, a note of outright hostility unheard for several decades. In 1756, Macaulay admitted, the king 'had merely anticipated a blow intended to destroy him'. But in 1740 he had committed in his attack on Silesia a 'great crime' and had shown 'gross perfidy' and 'selfish rapacity'.[5] Frederick de-

---

[1] Raumer, *Contributions to Modern History*, pp. 66, 76; Ranke, *Memoirs of the House of Brandenburg*, ii. 414.

[2] *Contributions to Modern History*, p. 66.

[3] Ranke, *Memoirs of the House of Brandenburg*, ii. 2; *Frederick the Great and his Times*, i, Preface, vi–vii, ix.

[4] Exemplified by Campbell's remark that 'Of all nations, I should be inclined to assign the first place, in point of morals, to the northern Germans'. (i, Preface, xxii–xxiii).

[5] T. B. Macaulay, *Critical and Historical Essays contributed to the Edinburgh Review* (London, 1883), pp. 821, 799–801.

served praise for his adherence to the great liberal principles of freedom of expression and religious toleration, and for his marvellous capacity for work. But these good qualities had been offset by his mania for directing and regulating far too many aspects of the lives of his subjects; and 'under this fair exterior he was a tyrant, suspicious, disdainful and malevolent'.[1] Though it was the work of a brilliant journalist rather than a historian this was a formidable indictment. Even in Britain, however, it barely modified the predominantly favourable picture of Frederick now well established in the mind of the reading public. In continental Europe, and above all in Germany, its influence was negligible. In the German world, almost simultaneously with the appearance of Macaulay's essay, the picture of *der alte Fritz*, the stern but just and kindly father of his people, was consolidated in the popular mind by the publication in 1840 of Franz Kugler's *Geschichte Friedrichs des Grossen*.[2] This book, the work of an art historian, had no claims to originality or penetration. But the attraction of its style, the telling (and often apocryphal) anecdotes upon which it relied heavily, and above all the five hundred illustrations by Adolf Menzel with which it was embellished, gave it immense popularity and made it one of the great myth-creating historical works of the century. Nearly two decades later another work of mythology, of a very different kind, began to appear with the first volumes of Thomas Carlyle's *History of Friedrich II, of Prussia, called Frederick the Great* (6 vols., London, 1858–65). This can be regarded as the last important pre-scientific book on Frederick to appear; the last, that is, to be written on a large scale but without any use of manuscript materials or even much of the printed documentation which was now beginning to be available in increasing quantities. It is a disappointing work which relies heavily on a fairly small number of secondary sources, notably Preuss,[3] and which seldom really comes to grips with the questions which have interested and divided historians of Frederick's reign.[4] Moreover the author

[1] *Critical and Historical Essays*, pp. 805, 807–8, 810.

[2] An English translation quickly appeared; Francis Kugler, *History of Frederick the Great* (London, 1844).

[3] The copy of Preuss's book in the London Library has numerous marginal notes by Carlyle which show how carefully he had worked through it; the *History of Friedrich II* contains a good many unacknowledged borrowings from the German historian.

[4] Note, for example, the lack of any effort at analysis of the issues involved in what Carlyle (*History of Friedrich II*, iv, Bk. XVII, chaps. i, iv) says about Frederick's position and actions in 1756.

himself clearly lost interest in the subject after a time with a resulting lack of proportion in the book: the first two volumes deal in great detail with the reign of Frederick William I and there is extensive coverage of the 1740s; but the last two decades of the reign are merely sketched. To Carlyle Frederick was admirable essentially because of his harsh realism, his willingness to face facts and shoulder the responsibilities of leadership. He could thus be used to illustrate and condemn the emptiness and superficiality of the eighteenth century which Carlyle so much hated, 'a Century which has no History and can have little or none . . . opulent in accumulated falsities'.[1]

The decades from the 1850s to the outbreak of the First World War saw the final triumph, in writing on Frederick the Great as in all aspects of historical writing in Germany, of the professional over the amateur. More and more the significant books and articles concerning him were the work of professional academics, the products of the great German universities which had done so much to create history as a modern academic discipline. Moreover they could now be based on an impressive mass of printed primary materials, itself the product of decades of devoted labour by scholars. From the later 1840s onwards a workmanlike edition of Frederick's own writings had been rapidly produced under the auspices of the Prussian government.[2] From the end of the 1870s a huge and beautifully-edited edition of his political correspondence was under way: by 1903 twenty-seven volumes had appeared, covering the most dramatic part of Frederick's reign down to the year 1768; the series had still reached only 1782 when it ended in 1939.[3] By the 1890s a mass of specialized materials on the administrative history of eighteenth-century Prussia was becoming available in the series *Acta Borussica: Behordenorganization*, of which fifteen volumes appeared between 1894 and 1936. Apart from this main series, moreover, a number of smaller ones were published: three volumes dealing with regulation of the grain trade (1896–1910); three on the history of the silk industry in 1892; four on the coinage (1904–13); and two on trade and tariff policy (1911–22). The historian writing in Wilhelmine Germany, therefore, had access to a large and growing fund of printed information of a kind which Ranke or Carlyle had hardly dreamed of.

[1] *History of Friedrich II*, i. 10–11.

[2] *Oeuvres de Frédéric le Grand* (30 vols., Berlin, 1846–56).

[3] *Politische Correspondenz Friedrichs des Grossen* (46 vols., Berlin, 1879–1939).

The 1860s, when the systematic publication of documents was only beginning, saw the appearance of perhaps the most bitter and effective attack ever made on Frederick, that launched by Onno Klopp in his *Der König Friedrich II von Preussen und seine Politik*.[1] Klopp, a Hanoverian driven to Vienna by the Prussian conquest of his native state in 1866, centred his criticisms on a single issue—that Frederick, by creating dualism in Germany, by establishing there a balance between Prussia and Austria, Hohenzollern and Habsburg, based on rivalry and distrust, had in the pursuit of his own selfish state and dynastic aims fatally weakened the entire German world. This was not a new charge: it had already been made by some of Frederick's romantic critics. But Klopp's formulation of it was more detailed, more systematic, and more bitter than anything seen before. The fatal attack on Silesia in 1740, he argued, was merely the first of a long series of wars of conquest waged by Prussia which differed in details but not in essentials. The results had been disastrous for the German people.

> Between Germans and French, Germans and Russians, there is a natural hostility. The antipathy which exists in our times between the Pomeranians, the Brandenburgers, and the inhabitants of the Steiermark and the Tyrol, is not a natural one but one manufactured, clearly fed by blood, by the blood which Frederick II allowed to flow. Let it be said clearly and in few words: King Frederick II not merely destroyed the thousand-year-old empire: he made peace between the Germans impossible.[2]

To this central accusation Klopp repeatedly and obsessively returned. In 1740 Frederick had had no just grievance against the emperor; and any he may have felt could not reasonably be extended to Maria Theresa, who did not succeed her father as ruler of the empire. The king had acted as he did simply because 'he wanted war, he wanted conquest, and this at the expense of the hereditary lands of the imperial house'. Even in 1756 it was unlikely that an aggressive war against Prussia would really have been launched (though Klopp could not deny that plans for such a war existed).[3]

From Frederick's aggressive ambitions there had flowed, in addition to the deep and unnecessary division which he inflicted on Germany, two other disastrous results. Prussia had been transformed

---

[1] The first edition appeared in 1860. My references are to the enlarged second edition (Schaffhausen, 1867).

[2] p. 128.

[3] pp. 122–4, 243 ff.

into a military despotism, its energies strained to the limit to produce the greatest possible armed strength and the resouces of each conquest used to achieve still further ones. 'Frederick', Klopp alleged, 'transformed the unfortunate country which fate had placed under his control, the state of Prussia as it came from his hands, into a machine for conquest'.[1] Even more serious, Frederick had been compelled, to achieve and retain his conquests, to rely to a shameful and dangerous extent on foreign help. In 1740 he had played France's game by attacking Austria. In 1744 by his invasion of Bohemia he had made it impossible for the Habsburgs to recover Alsace from France. Above all he had neglected his duty as a German to ally with them against the threat posed by Russia, 'this barbaric empire'. On the contrary he had helped the growth of Russian power, notably by the treaty of alliance which he signed with Catherine II in 1764. 'There is, perhaps,' wrote Klopp, 'particularly for us Germans, no more shameful treaty than this', for 'the day of the signature of this treaty is, for those Germans who belong to the state of the Hohenzollerns, the beginning of their enduring vassalage to Russia.'[2]

Long-winded, repetitive, deeply biased, this was none the less a book of importance. It can be regarded as the first systematic and powerful statement of the idea of the inherent wickedness of the Prussian state as it had existed from the eighteenth century onwards, a wickedness which Klopp, very unhistorically, traced entirely to the imprint of a single evil genius, Frederick II. Some of the accusations the book made had much force. Frederick had indeed consolidated, though he did not create, the essentially military character of the Prussian state. He had undoubtedly during much of his reign attached overriding importance, so far as foreign policy was concerned, to relations with Russia. Klopp, for all his exaggerations and lack of balance, had set in motion a controversy which was to last down to the present day. There was no lack of replies to his arguments from the Prussian side. Even before he published his book Ludwig Hausser had presented the other side of the coin by claiming that the Habsburgs had always preferred their selfish dynastic interests to those of Germany as a whole. If the anti-Prussian alliances of 1756–7 had been successful, he argued, Russia would have taken East Prussia, Sweden Pomerania, and France territory on the left bank of the Rhine; in fact there would have been a repetition for Germany

[1] pp. 403–4.
[2] pp. 132, 157, 150, 322, 329.

of the disastrous territorial losses of the peace of 1648.[1] Later, during the Bismarckian period, such exponents of the official viewpoint as Treitschke, Droysen, and Sybel also hastened to refute Klopp's accusations. Some of these refutations took merely the form of allegations that, far from creating dualism in Germany, Frederick had clarified and given necessary political expression to a division which had existed since the Reformation. Thus Treitschke who, though a Saxon by birth, expressed more uncompromisingly than any other writer the extreme Prussian viewpoint, claimed that 'it was not Frederick who created German dualism, though he was reproached with this by his contemporaries and by posterity; dualism had existed since the days of Charles V'.[2] More important, it could be argued that Austria had never been a truly German power and therefore had not, under Maria Theresa or Joseph, been entitled to any dominant influence in German affairs. The weakening of her influence in the German world, in other words, and its replacement by that of Prussia, had been a progressive and necessary step. Thus Treitschke compared the healthy concentration of the Hohenzollerns on the building up of their own state with the un-German attitude of the Habsburgs, 'whose statesmanship was wholly concerned with European questions'. In the same way almost a generation later Koser lauded the Prussia of Frederick II as an exclusively German great power such as had not hitherto existed in Europe. Austria, weighed down by her non-German territories, could never have become such a power. Although, he admitted, the Hohenzollerns had been unconscious of any German national mission, each of their annexations had none the less served not merely their own dynastic interests but also the interests of the German people: 'what Prussia gained Germany won'.[3] It was a boon for Germany that neither the Bavarian attempt to conquer Bohemia in 1741 nor the Austrian efforts to absorb Bavaria in 1745 had borne fruit. If either had succeeded the result would have been the setting up in south Germany of a really powerful Catholic state: this would have estab-

---

[1] *Deutsche Geschichte vom Tode Friedrichs des Grossen bis zur Gründung des deutschen Bundes* (2nd edn., Berlin, 1859–60), i. 61. Klopp replied to these arguments in a pamphlet, *Offener Brief an den Herrn. Professor Hausser in Heidelberg betreffend die Ansichten über den Friedrich II von Preussen* (Hanover, 1862).

[2] *Treitschke's History of Germany in the Nineteenth Century* (London, 1915), i. 53. The German original was published in 1879.

[3] Treitschke, *History of Germany*, i. 53; Koser, *Geschichte Friedrichs des Grossen*, ii. 555.

lished dualism of a truly dangerous kind and made the unity eventually achieved in 1871 far more difficult to accomplish.[1] Through every discussion from the Prussian standpoint of the question of dualism there sounds the now deep-rooted assumption that the victory of German nationalism and of progress on every level was indissolubly bound up with Protestantism, that Catholicism by its very nature was a retrograde force opposing national unity.

These defences of Prussia, however, did not end the controversy which Klopp had opened with such bitterness. It became more and more impossible to deny Frederick's contemptuous indifference towards any idea of German nationality and his willingness to destroy the unity of the German world whenever it seemed in his interests to do so. In 1920, for example, G. B. Volz, one of the greatest experts on the period, pointed out that in 1759, at the most critical moment of the Severn Years War, he had seriously contemplated dividing the whole of north Germany between Prussia and Hanover and making both kingdoms completely independent of the remains of the old Reich.[2] Moreover Klopp's arguments appealed powerfully to the dislike of Prussia which was widespread in south Germany and the Austrian provinces in the later nineteenth century; and the collapse of 1918 gave this dislike freer play than it had enjoyed during the last fifty years. In 1925, therefore, Werner Hegemann was able to centre the most bitter personal attack on Frederick which had appeared for over two generations very largely around the charge that he had, by his rivalry with Austria, deprived Germany for ever of the chance of dominating central Europe and thus becoming a true world power on the scale of Britain, the U.S.A., or Russia. Hegemann referred contemptuously to 'that Prussian little Germany which Frederick II and Bismarck were at pains to substitute for the great Central European Empire of Prince Eugene and Maria Theresa', and claimed with bitterness that 'all the barbarous German squabbles started by Frederick had only one common significance for all Germans, namely, that they weakened Germany and secured her defeat in the struggle for the lost provinces and for a "place in the sun" at the very moment when these might have been secured'. In the First World War Germany had tried suddenly and without preparation to undo the disastrous effects of the period 1740–1914 and to create

---

[1] R. Koser, 'Staat und Gesellschaft zur Hoheit der Absolutismus', p. 324.

[2] 'Friedrich den Grossens Plan einer Losreissung Preussens von Deutschland', *Historische Zeitschrift*, cxxii (1920) 267 ff.

a German-dominated central Europe; but by then it was too late and such schemes were mere romanticism.[1] Hegemann was a littérateur, and a violently biased though well-informed one. However ten years later his arguments were repeated, in a less polemical tone but with little fundamental alteration, by Ritter von Srbik in the last great academic statement of the *Grossdeutsch* indictment of Frederick as the splitter and weakener of Germany. In some ways he went further than Klopp, arguing that the loss of Silesia by the Habsburgs had been quite disastrous for the future of the German people. It had fatally weakened the German colonizing effort in central Europe (since this depended on keeping the Habsburgs strong), and had handed over Bohemia, the essential link between north-east and south-east Germany, to a future threatened by Slav domination. By weakening the Germans of Austria against the Poles and Hungarians it had undermined the entire movement of German cultural penetration to the south-east. Moreover Frederick, by creating in Prussia a *Willenstaat* dominated by an incessant drive to power, had helped to form a 'north-German–Prussian type of man' characterized by a hard mechanical outlook and rigid ideas of duty: in this way a Prussian spirit different from the German national one had been generated and a profound psychological rift established in Germany. And all this had been done with foreign help, since for Prussia 'foreign subsidies, alliances with France at the expense of the Reich, a policy of balance between Habsburg and Bourbon and changes of course dictated by self-interest, bore rich fruit'.[2]

Inevitably efforts were made to escape from the dilemma which Klopp had so harshly posed; that of choosing, in effect, between the greatness of Prussia and the long-term interests of Germany as a whole. A decade after the publication of Klopp's book, Ranke in one of his later works argued that throughout his reign Frederick had sought to reconcile the interests of the German states in general with those of Prussia. (Though he admitted that in this he had been unsuccessful, at least until the formation at the end of his reign of the *Fürstenbund* of 1785, the grouping of German rulers formed under Prussian leadership to resist the ambitions of the Emperor Joseph II).[3]

[1] W. Hegemann, *Frederick the Great* (London, 1929), pp. 317, 508–9. This English translation is a slightly shortened version of the German original of 1925.

[2] H. Ritter von Srbik, *Deutsche Einheit. Idee und Wirklichkeit vom Heiligen Reich bis Königgrätz* (4 vols., Munich, 1935–42), i. 102, 107, 95.

[3] *Die deutschen Mächte und der Fürstenbund* (2 vols., Leipzig, 1871–2), i, chap. xiii.

Efforts to show that Frederick's achievement transcended the boundaries of Prussia and had a meaning for all Germans have indeed continued down to the present day. Had he not provided Germany with a hero around whom national feeling could crystallize? Had he not, in 1740, promised Maria Theresa support against all her enemies in return for the cession of most of Silesia, and in 1756 hoped for a quick victory which would have placed the armies of all the German states under his control to be used in the expulsion of foreign forces from the empire? More seriously, had not his regime in Prussia acted as a model to be copied by the other German rulers (notably Joseph II himself) and did not this endow his work with a truly *gesamtdeutsch* aspect by the impetus it gave to the development of a great, centralized, German-dominated Habsburg empire?[1] But the dilemma remained; the question of the meaning and justification of Frederick's work was still open to more than one answer.

The decades from the German victory over France in 1871 to the First World War nevertheless saw the historical reputation of Frederick at its height. In Germany he seemed to the majority of scholars more clearly marked than ever before as the greatest figure in the process which had culminated in the foundation of the second Reich in 1871. An official orthodoxy of admiration for him was now well established, so that even the best-informed academic criticism was likely to arouse deep resentment. Thus when in 1894 Max Lehmann in his *Friedrich der Grosse und der Ursprung des siebenjährigen Krieges* suggested that in 1756 Frederick had merely used the hostility to him of Maria Theresa and the Empress Elizabeth as a pretext for an effort to gain territory in Saxony, that his policies in that year had not been in any deep sense defensive at all, he aroused a storm of passionate hostility.[2] In intellectual terms the criticisms made of Lehmann were often justified; but the bitterness of the controversy which he aroused shows how deeply the tendency to glorify Frederick had now taken root in Prussian academic circles. Nor was this tendency to be seen only in Germany. In particular French interest in Frederick had been greatly increased by the

[1] There is a useful summary of these arguments in W. Schüssler, 'Friedrich der Grosse in gesamtdeutscher Schau', in his collection of essays *Deutsche Einheit und gesamtdeutsche Geschichtsbetrachtung* (Stuttgart, 1937), pp. 53–68.

[2] On this controversy see Sir H. Butterfield, *The Reconstruction of an Historical Episode: The History of the Enquiry into the Origins of the Seven Years War* (Glasgow, 1951), pp. 19–21, 41–2.

events of 1870–1: there was now in Paris a much more active desire than ever before to learn about the new great power which had overwhelmed France, and a deep current of grudging admiration for her. It is significant that one of the most wholeheartedly favourable pictures of Frederick's personality, which stressed in a romantic way the ambiguities, complexities, and contradictions of this 'tormented soul', should have been the work of a Frenchman.[1]

A number of elements made up this highly favourable view of Frederick, none of them new but all of them now more forcibly expressed and perhaps more widely accepted than ever before. The first of these was the belief that Prussia expressed a higher and purer morality than her opponents, that Frederick's victories owed little to superior material resources but were the fruit of effort, self-sacrifice, a sense of duty, which the Austrians and the French simply could not match. They were seen, in a word, as the victories of a superior state and society over inferior ones. Sometimes this attitude took what can fairly be called a racialist form, as in Treitschke's claims that 'the vigorous will of the North German tribes gave them from the earliest times a superiority in state-constructive energy over the softer and wealthier population of High Germany', and that 'the charms of sin were not felt by the heavy North German nature'.[2] Often it was expressed in a more sophisticated way, in praise of the deep sense of duty, the rational and freely accepted discipline, the combination of wide tolerance with effective power, which were held to have made the Prussian state the greatest achievement of the German Enlightenment.[3] Occasionally it took the form of regarding Frederick's policies as in some sense the practical embodiment of Kant's categorical imperative.[4] Frequently it was accompanied by comparisons of Prussian virtue with the 'fragile morality of the Parisian enlightenment' and the 'revolutionary frivolity' of the French.[5]

---

[1] L. Paul-Dubois, *Frédéric le Grand d'après sa correspondance politique* (Paris, 1903), especially pp. 9 ff., 319–28.

[2] *History of Germany*, i. 28, 42.

[3] W. Dilthey, 'Friedrich der Grosse und die deutsche Aufklärung', *Gesammelte Schriften*, iii (Leipzig–Berlin, 1927), 134. This essay, written about the beginning of the twentieth century, is the best statement of the case for regarding Frederick as a figure of real intellectual importance in his own right.

[4] O. Hintze, 'Geist und Epochen der preussischen Geschichte', *Gesammelte Abhandlungen*, iii (Göttingen, 1967), 20. This essay was first published in 1903.

[5] Treitschke, *History of Germany*, i. 95; Dilthey, 'Friedrich der Grosse', 201.

The successes of Prussia were glorified not merely as the victories of a superior morality but also as those of a higher form of religion. The deep sense of Protestant superiority to Catholicism, a feeling clearly visible in such towering figures as Hegel and Ranke, which had been from the beginning one of the most profound sources of sympathy with Frederick, was as strong as ever in Germany under Bismarck and William II. Thus Treitschke stressed that the struggle of Prussia for survival in the Seven Years War had also been a great religious struggle for the defence of Protestantism and rejoiced that 'the establishment of the Protestant–German great power was the most serious reverse that the Roman See had experienced since the rise of Martin Luther.'[1] Though few writers expressed this attitude so uncompromisingly, it was one which underlay a great deal of the favourable comment of the period on Frederick. It faithfully expressed the outlook of the ruling forces, Prussian and Protestant, in the newly-unified and in many ways still insecure Germany of the decades after 1871.

The most notable new departure of the Bismarckian and Wilhelmine period, however, was the increased emphasis which it placed on Frederick as a German patriot. Persistently, and sometimes with a cavalier disregard for the facts of history, the belief was propagated that he had been a German nationalist, the protector of German rights against the pressures and hostility of the outside world, and that the German people even in his own day had recognized him as such. More and more he was presented as a purely German figure thinking in German terms and working within a German frame of reference. The cosmopolitanism of his outlook, the European aspects of his achievements, tended to be systematically played down in favour of a concentration on his alleged German patriotism. The contrast between the very detailed treatment of Frederick's foreign and military policies in J. G. Droysen's *Geschichte der preussischen Politik* (5 vols., Berlin, 1855–76) and that offered earlier by Ranke is instructive here. Ranke had seen the king's victories in a European context. He had attributed them, quite correctly, in large part to the structure of international relations in Europe in the mid-eighteenth century and Frederick's ability to use this to his advantage. Droysen on the other hand saw them much more in terms merely of the German situation and the workings of Frederick's strong and independent will. Significantly, Ranke began

[1] *History of Germany*, i. 69–70.

his study of the subject by extensive research in Paris while Droysen never worked outside Prussia.[1] Treitschke was again the most whole-hearted and uncritical exponent of this nationalist view of Frederick, as in his quite unhistorical insistence that 'his soul was animated by a vigorous national pride' and that 'he is never for a moment befooled by the thought of cutting his own state loose from the fallen German community'.[2] Beliefs of this kind, however unjustified, were by the 1870s part of the mental furniture of many Germans. By one of the ironies of history this ruler, throughout his life unconscious of German national feeling and contemptuous of German culture, had become and was long to remain a symbol of German national unity and pride. In particular the overwhelming defeat he had inflicted on the French at Leuthen in 1757 became a focus for such feelings, so that in 1940 the exiled William II, telegraphing congratulations to Hitler on the collapse of France, could write that 'in all German hearts there sounds the Leuthen chorale'.[3]

Not all German judgements of Frederick were completely favour-able during these decades when his reputation in general was at its apogee. In academic circles Lehmann criticized his aggressive expansionism;[4] while in the early years of the twentieth century Otto Hintze, now emerging as the greatest all-round expert on the history of Prussia, declared forcibly that 'the German idea as a constructive principle of Prussian history is . . . an illusion'. He also made more clearly than any earlier academic commentator a potentially very important point; the extent to which Frederick had favoured and protected the Prussian nobility. In the eagerness with which he co-operated with and relied on the great Prussian landowners Hintze correctly saw one of the really fundamental differences between Frederick and his father.[5] Almost simultaneously the same issue became one of the main bases for a scathing attack on Frederick from the Marxist standpoint by Franz Mehring in his article 'Ein aufgeklärter Despot?' which appeared in the Social Democrat periodical *Neue Zeit* in 1912.[6] Unlike other eighteenth-century

[1] Bussmann, 'Friedrich der Grosse im Wandel des europäischen Urteils', pp. 381-2.
[2] *History of Germany*, i. 61-2.
[3] R. Augstein, *Preussens Friedrich und die Deutschen* (Frankfurt, 1968), p. 8.
[4] See p. 146 above.
[5] 'Geist und Epochen der preussischen Geschichte', *Gesammelte Abhand-lungen*, iii. 2; 'Die Hohenzollern und der Adel', *ibid.*, 45-50.
[6] English translation in P. Paret (ed.), *Frederick the Great: a Profile*, pp. 221-8.

despots, Mehring accused, Frederick had not followed a progressive policy of strengthening the towns and the middle classes against the landowning nobility. Instead he had placed the whole machinery of the Prussian state in the hands of the Junkers, sacrificing to them the interests of both towns and peasants. Nor had he the slightest tincture of German national feeling: he would happily have surrendered East Prussia to Russia and the Rhenish possessions to France if in return he could have annexed Saxony, the territory which he wanted above all. During the Seven Years War he had been the vassal of England; after it he became the satellite of Russia, while the collapse of 1806 showed how little solidity or permanence his work had. In the attention it gave to the social basis of his regime this article illustrated the development of a potentially fruitful new approach to the study of Frederick and his reign. Yet mixed with its bitter condemnation was a strain of grudging admiration for the king's abilities, realism, and capacity for work. 'Compared to the degenerate pack of princes of his time,' Mehring admitted, 'Frederick was a real man . . . At least he carried on his evil work with a degree of energy and seriousness that was alien to his peers.'[1]

During the half-century before 1918 any critic of Frederick the Great, Marxist or otherwise, was very clearly swimming against the academic tide, in Germany and indeed in Europe as a whole. After 1918 this position changed. The collapse of the imperial structure and the end of the Hohenzollern dynasty threw, as never before since 1806, severe doubt on the value of Frederick's achievement. The temporary strengthening of liberalism and radicalism of all kinds under the Weimar regime encouraged a much more critical view of his policies than had been normal during most of the nineteenth century. After 1945, in a Germany overshadowed by a much greater catastrophe, criticism became still more searching and forthright. A good deal of this criticism differed noticeably in tone, however, from that of the nineteenth and early twentieth centuries. Klopp had adopted towards Frederick a consistent attitude of moral

---

[1] p. 225. In an earlier work, *Die Lessinglegende* (Stuttgart, 1893) Mehring had presented a Marxist picture of Frederick as governed in what he achieved by the workings of economic factors. He had praised, however, the intelligence and realism the king showed in grasping the limitations of his own position (pp. 108, 127). Earlier still, in the 1850s and 1860s, Ferdinand Lassalle, one of the founders of German socialism, had shown some sympathy with Frederick for roughly the same reasons (Bussmann, 'Friedrich der Grosse im Wandel des europäischen Urteils', pp. 394–5).

outrage; but he had not been concerned with the failings of his daily and personal life. He had hated Frederick's policies but he had been less interested in his weaknesses as an individual. Mehring had attacked bitterly the social basis and meaning of Frederick's regime but had respected his personal qualities. After 1918, and still more after 1945, it became possible as never before not merely to attack Frederick as a historical phenomenon but to dislike and even despise him as a man, to call attention as never before to everything which could be urged against him as an individual. Thus in the 1920s Hegemann devoted a large part of a substantial book to niggling and often unfair personal critcism of the king, weakening his case unnecessarily by a blatant refusal to credit him with any good qualities at all. Frederick had not, Hegemann alleged, worked nearly so hard or so conscientiously as nationalist historians had claimed; the conventional picture of a monarch driven by an overmastering sense of duty to toil incessantly for the good of his people was a ridiculous caricature. On the contrary he had been lazy, addicted to sleeping much longer than was really necessary. Far from being frugal in his personal expenditure, as the traditional pieties of German historiography had proclaimed, he had been more extravagant than Louis XV or Madame de Pompadour. His taste in architecture had been bad: even his military abilities had been grossly overrated.[1] Over forty years later, in the best book hitherto written on Frederick from a critical point of view, Rudolf Augstein presented another highly unfavourable sketch of his character, though one much more balanced and sophisticated than Hegemann's. Augstein does not deny that Frederick had a profound sense of duty, or that he was the most witty and ingenious (*geistreichste*) ruler of his age. But his character lacked nobility. Though he allowed free criticism in religious matters and even in intellectual life he forbade any of his regime or his methods of government. He despised his subordinates and tried in a petty and small-minded way to deny them credit which they deserved. If he was frugal in some respects he was notably extravagant in others.[2] Before 1918, with the Prussian monarchy an apparently viable and indeed growing force, criticisms of this kind had been muted and unimportant. During the last two generations, as it has become increasingly fashionable to see that monarchy as a disastrous dead-end in German history, the personal weaknesses of

---

[1] *Frederick the Great*, pp. 354 ff., 377 ff., 383 ff., 281 ff.
[2] *Preussens Friedrich und die Deutschen*, Chap. 9, *passim*.

its greatest figure have become a more and more tempting target for both academic and popular writers.

The substance as distinct from the tone of what historians have said about Frederick, the issues which they have considered important, have also changed over the last half-century. Any pretence that he was a German nationalist or that he thought in terms of anything but Prussian state and Hohenzollern dynastic interests has now been almost completely given up. The most balanced and judicious summing-up of his reign to date, the work of an author in many ways conservative and nationalist, concludes flatly that 'his policies show not the slightest awareness of national responsibility'.[1] Frederick's critics have driven this point home more firmly than ever before, making with renewed vigour and the support of abundant historical detail all the points which Klopp raised so passionately in the 1860s— Frederick's Prussian and dynastic aggressiveness and expansionism; his hostility to Austria and the old imperial structure; his willingness, even anxiety, to ally with foreign powers for his own purposes.[2]

More novel has been the marked growth of interest in the structure and workings of the Frederician regime in Prussia. The Prussian administration, in some respects the army, and the social foundations and implications of both, have been studied with a thoroughness to which no earlier period could aspire. The most prominent writers of the nineteenth century, Ranke, Raumer, Carlyle, Droysen, Klopp, even Koser, were all concerned with foreign policy and war rather than with the administration, finances, or society of Prussia. By the first years of the twentieth century this position was beginning to change, notably through the work of Otto Hintze. Only within the last few decades, however, have Frederick's accomplishments within Prussia received as much attention as those in war and foreign policy. Most of this work has been done in an intellectual atmosphere unfriendly to the whole ethos of the Prussian state. It is not surprising, therefore, that it has produced in general hostile verdicts. Criticism has focused in particular on two points; the unwieldiness of the over-centralized administrative machine which Frederick inherited and developed, and the extent to which it was dominated by military considerations and a military psychology. The first of these was brought home to the English-speaking public in a series of important

[3] G. Ritter, *Frederick the Great, an Historical Profile* (London, 1968), p. 47.
[2] The best statement of this now orthodox argument is in Augstein, *Preussens Friedrich*, chap. i, *passim*.

articles published by W. L. Dorn in 1931–2.[1] These illustrated very effectively the extent to which, in the later part of his reign, Frederick was unable to control effectively a more and more complex mechanism of government in which all real power of decision was reserved, at least in theory, to the king himself. In the writing of the last generation or more it has become in fact almost a truism to point out that in his later years difficulties and unpleasant truths were kept from Frederick by his subordinates, and that his desire to control in person every lever of government and his liking for quick decisions tended in practice to throw power increasingly into the hands of his ministers and the heads of the provincial chambers.[2] The picture drawn by the nationalist historians of the nineteenth century was that of the great monarch controlling in person, through incessant work and the exercise of a brilliant intelligence, all the threads of government. This tends today increasingly to be replaced by the perhaps equally misleading one of an irascible and opinionated old man struggling to dominate a machine which by his death was already in many respects outdated.

Augstein in particular has insisted on the ineffectiveness of many of the king's policies, especially in the economic field. His efforts in this area were ignorant and disorganized; he had no Colbert and would not have known what to do with one if he had possessed such a minister, for he wished simply 'to drill the economy like a sergeant'.[3] His excise experiment of 1766 merely created a new and expensive bureaucracy without achieving any economic or social purpose. In Silesia, which was governed by a minister personally responsible to Frederick and of which the king was therefore in a sense in personal control, trade and the population of many of the towns actually declined after the Seven Years War. Prussia under Frederick's rule suffered from 'a backward, autocratic and sergeant-

---

[1] 'The Prussian Bureaucracy in the Eighteenth Century', *Political Science Quarterly*, xlvi (1931), 403–23, xlvii (1932), 75–94, 259–73.

[2] e.g. H. Haussherr, *Verwaltungseinheit und Ressorttrennung vom Ende des 17. bis zum Beginn des 19. Jahrhunderts* (Berlin, 1953), p. 142. This short book is an interesting and ambitious attempt at a comparative study of the administrative systems of the main European states in the eighteenth century. W. Merteneit, *Die friedericianische Verwaltung in Ostpreussen* (Heidelberg, 1958), pp. 183–4, shows how after the Seven Years War Frederick's increasing tendency to deal directly with the provincial chambers and their presidents tended to give the latter greater authority and to weaken the General Directory, the main institution of central government.

[3] *Preussens Friedrich*, pp. 224, 234.

like (*feldwebelhaft*) economy'.[1] So sweeping and well-argued a rejection of Frederick's economic policies would hardly have been possible before 1918 or even 1945. The nineteenth century had indeed seen some criticism of this aspect of his reign, for example in liberal disagreement with his efforts to control the grain trade in Prussia.[2] But its verdict on Frederick's economic policies, as on everything he did, had been generally favourable. In particular the protectionist and 'state-building' side of his efforts to foster industry in Prussia, and to develop its trade and equip it with new financial institutions, had had some affinity with the current of protectionism which was running strongly in Germany from the 1870s onwards. They had therefore won the warm approval of such important neo-mercantilist writers as Gustav Schmoller.[3] In this respect as in others the cataclysms of the twentieth century drastically altered the position and opened the way to criticism of Frederick of a violence hitherto hardly thinkable.

The most emotionally weighted of all the charges laid against the king after 1945, however, was that his rule, in its military and aggressive character, had in some sense anticipated the Nazi regime.[4] Few writers went so far as this; but the revulsion against the Prussian military tradition which followed the Second World War led to unprecedentedly severe criticism of Frederick's army and its influence. To Augstein it was a mere *Schiessmaschine* (shooting-machine) to be set in motion by the king, a horde of slaves officered by the territorial nobility for whom alone army service could be considered an honour. Its ferocious discipline made it a forerunner of the Nazi concentration camps; and its political and social power had turned Prussia into a *Mamelucken-Staat*.[5] Such criticisms are too extreme to be acceptable to most historians; but there is no doubt that the moral stature of Frederick's Prussia has declined sharply by comparison with the esteem which it enjoyed before 1918. Today it is largely stripped of the progressive, rational, and 'enlightened' meaning with which it was

[1] *Preussens Friedrich*, pp. 240–7, 233, 264, and Chap. 7 *passim*.

[2] e.g. M. Philippson, *Geschichte des preussischen Staatswesens* (Leipzig, 1880), i. 19.

[3] See the comments in his *The Mercantile System and its Historical Significance* (London, 1895). The book is an English translation of a chapter in his *Studien über die wirtschaftliche Politik Friedrichs des Grossen* (1884).

[4] Augstein, *Preussens Friedrich*, pp. 147, 350, 353.

[5] *Preussens Friedrich*, pp. 128, 130–1, 145. For a more moderate statement of the case against Frederick's militarism see L. Dehio, 'Um den deutschen Militarismus', *Historische Zeitschrift*, clxxx (1955), especially 47 ff., 57.

formerly clothed. To many educated Germans it now means little more than the cult of order carried to extremes and complete submission to the state.[1]

Inevitably, sharp criticism has prodcued a reaction. Attacks on Frederick and the tradition he did so much to create have stimulated energetic defences of the threatened hero and his work. The existence of any link between the Prussia of Frederick, a *Rechtstaat*, and the Germany of Hitler or even that of Bismarck has been indignantly denied.[2] Frederick, it has been argued, was not a militarist, and his greatness rests not merely on his achievements in the field but rather on his entire political and cultural activity, inside and outside Prussia, on 'his representation of a culture'. The year 1740 did not, as some historians have misleadingly tried to argue, determine the tone of the whole of Prussia's future history: in any case in that year Frederick offered a less dangerous threat to the Reich than the elector of Bavaria had nearly four decades earlier by siding with the French at the beginning of the Spanish Succession war.[3] Even the view of Prussia as dominated by the landed nobility has been challenged by the argument that since in the later eighteenth century all schoolmasters and most holders of judicial posts were members of the middle classes the bureaucracy was really the creation of an enlightened bourgeoisie.[4]

Two modern writers above all have avoided the extremes of unfair condemnation and uncritical admiration. They are Friedrich Meinecke, in his study of Frederick's political ideas, *Machiavellism* (New Haven, 1957) and Gerhard Ritter in his wider-ranging *Frederick the Great: An Historical Profile*.[5] The two writers achieve a high measure of agreement. Both see Frederick as dominated throughout his life by considerations of rational state necessity.

[1] See the comments of G. Wytrzens, 'Sur la sémantique de l'Aufklärung en Allemange, en Autriche et dans les pays slaves non Russes' in P. Francastel (ed.), *Utopie et institutions au XVIIIe siècle* (Paris, 1963), p. 316.

[2] W. Hubatsch, *Das Problem der Staatsräson bei Friedrich den Grossen* (Göttingen, 1956), pp. 27–31; Schoeps, *Die Ehre Preussens*, pp. 6–7. The second is significant as coming from the pen of a Jew whose father died in a Nazi concentration camp.

[3] Hubatsch, *Das Problem der Staatsräson*, pp. 28–9, 33. This apologetic current in recent German writing is interestingly summarized in P.-P. Sagave, 'Les Historiens allemands et la réhabilitation de la Prusse', *Annales*, 15e année (1960), No. 1, 130–50.

[4] H. Brunschwig, 'Propos sur le prussianisme', *Annales*, 3e année (1948), 16–20.

[5] The first of these was originally published in German in 1924. The second is based on a course of lectures given at the University of Freiburg in 1933–4.

Both see him as continually torn between the ideal of enlightened and humanitarian rule and the need for state power; and they agree that in the last analysis the second of these always took precedence in his mind. But both insist that the ideals of the Enlightenment never ceased to have real meaning and to hold deep attractions for Frederick. 'Regardless of the enormous sacrifices the Prussian military state demanded of its inhabitants, and regardless of the frequently harsh nature of its institutions,' writes Ritter, 'in the eyes of its architect it was not simply a despotism. On the contrary, the state was turned consciously onto the path of justice and culture.'[1] In spite of much bitter criticism of Frederick during recent decades this is the fairest and most convincing verdict at present available.

One last point must be made. This is the extent to which recent writing has illustrated and detailed the debt of Frederick to his father and the degree to which his achievements were made possible by the earlier adminstrative and financial struggles of Frederick William I. Though contemporaries were of course aware of Frederick William's efforts to build up a great army and accumulate a large reserve of cash[2] he was during the eighteenth century completely eclipsed by his brilliant son. Not until the 1830s did a substantial biography of him and an attempt at a systematic account of his administration of Prussia appear.[3] During the nineteenth century it was rare to find among historians any adequate appreciation of the significance of Frederick William I in the history of the Prussian monarchy. To Macaulay he was merely an avaricious brute, an uncultured military maniac; to Kugler no more than a half-crazy tyrant.[4] Carlyle, though he devoted the first two volumes of his book on Frederick the Great to the reign of his father, was concerned only with foreign policy and the relations between father and son; the constructive importance of Frederick William I largely escaped him. Even Ranke succumbed to the same temptation; though he realized that 'in finance, and his whole internal administration, Frederick followed . . . in the footsteps of his father'[5] he gave his readers no

---

[1] *Frederick the Great*, pp. 71–2.

[2] e.g. Mr. de M. . . . (i.e. Eleazar de Mauvillon), *Histoire de Frédéric Guillaume I, Roi de Prusse* (Amsterdam–Leipzig, 1741), ii. 443–4.

[3] F. Förster, *Friedrich Wilhelm I, König von Preussen* (3 vols., Potsdam, 1834–5). The second half of vol. ii gives much information on his administrative policies.

[4] *Critical and Historical Essays*, pp. 792–4; *History of Frederick the Great*, pp. 38–87.

[5] *Memoirs of the House of Brandenburg*, iii. 460.

adequate account of what the father had achieved in the reorganization and strengthening of Prussia. Treitschke grasped better than
many of his predecessors the fundamental importance of Frederick
William, and typically justified his boorishness on the grounds
that 'The firm and manly discipline of a fighting and industrious
people was of greater importance for Prussia's high destiny than were
the premature blossoms of art and science.' He also saw clearly how
great was the debt of son to father so far as the administration of
Prussia was concerned.[1] Yet it was not until 1941 that Carl Hinrichs
produced, in his *Friedrich Wilhelm I, König in Preussen . . . Jugend
und Aufstieg*, a biography of the 'sergeant-king' worthy of its subject's
historical magnitude. Sensitive and perceptive, Hinrichs's book is
one of the finest studies of any eighteenth-century ruler. Its appearance
placed its brutal, deeply unhappy, and yet strangely impressive subject
in a more favourable light than ever before. Almost simultaneously
Frederick William gained sympathy and understanding in a different
and more limited way through the publication of the pictures which
he painted in an effort to assuage his neurotic unhappiness.[2] After
the Second World War there appeared at last an adequate account
of his work as an administrator and the most important single
creator of the Prussian monarchy in the form in which it existed for
the rest of its life.[3]

Historians have been slow to do justice to Frederick William. But
the justice, though belated, has been full. One of the most important
contributions of the last forty years to the study of Frederick the
Great has been simply the achievement for the first time of an
adequate realization of the extent of his debt to his father.

RUSSIA: CATHERINE THE GREAT

Russian historiography, one of the most distinctive in Europe,
differs strikingly from that of the western states, notably Great
Britain, in its lack of any strong biographical tradition. In most other
directions it is well developed. It shows great strength in the field of
intellectual history and is almost embarrassingly rich in that of social
and economic history. In particular it has generated an enormous

[1] *History of Germany*, i. 44, 82, 88.

[2] J. Klepper (ed.), *In tormentis pinxit: Briefe und Bilder des Soldatenkönigs*
(Stuttgart–Berlin, 1938), reproduces twenty-six of these paintings.

[3] R. A. Dorwart, *The Administrative Reforms of Frederick William I of Prussia*
(Cambridge, Mass., 1953).

literature on the peasant question, the predominant preoccupation
of Russian historians during much of the nineteenth century.
But these resources show up all the more strikingly its relative poverty
in large-scale and scholarly biographies of individuals—statesmen,
diplomats, to perhaps a lesser extent soldiers, and certainly rulers.
When one remembers the extent to which real powers of decision at
the highest level in Russia remained, until the second half of the nineteenth century if not down to 1917, in the hands of the monarchs,
the extent to which their lives and personalities have been neglected
by historians is remarkable.

In no case is this more obvious than that of Catherine II. The
literature on her produced by the writers of her own day and of
succeeding generations is very extensive. But the proportion of gold
to dross is low. On the one hand there is a mass of writing about
her which is anecdotal, gossipy, sensational, and sometimes
scandalous. On the other there are serious studies of important
aspects of her reign—the extension and intensification of serfdom
in Russia, the territorial growth of the empire, the efforts at administrative improvement—in which the empress herself is relegated to
the background and often plays scarcely any role at all. The contrast
with Prussia or even with the Habsburg empire is striking. The
history of the former in the eighteenth century must be written
largely in terms of Frederick William I and Frederick II: that of
the latter, during his relatively short period of unchecked personal
power, is inseparable from the personality and aspirations of Joseph
II. In Russia, so much larger geographically and with problems in
many ways so much more deep-rooted and intractable, the historian
must often feel that impersonal forces and sheer physical difficulties
dominate the situation so completely that no ruler, however able
and allegedly all-powerful, can do more than affect the surface of
events. Nevertheless the wealth of Russian historical writing,
especially in its palmy days between the 1860s and the First World
War, is so great that adequate material is available for a study of
changing ideas of the personality and achievements of Catherine.[1]

[1] The history of historical writing in Russia has been a considerable preoccupation of the country's scholars. Some of the more obvious treatments are:
V. O. Klyuchevskii, 'Lektsii po russkoi istoriografii', in his *Sochineniya*, viii
(Moscow, 1958), 396–453 (this essay, written in 1892, deals with Russian historical
writing in the eighteenth century); P. N. Milyukov, *Glavnye techeniya russkoi
istoricheskoi mysli* (2nd edn., Moscow, 1898); M. N. Tikhomirov, 'Russkaya
istoriografiya XVIII veka', *Voprosy Istorii*, 1948, No. 2, 94–9; N. L. Rubinshtein,

Inevitably, much of this writing shows a tendency to read history backwards, to interpret the past in the light of the assumptions and preoccupations of the present. An eighteenth-century pamphleteer or collector of gossip, a nineteenth-century populist, a Soviet academician—all these will provide different pictures and strikingly different evaluations of the empress and her work. These variations, however, have been less complete and less drastic than in the treatment of Frederick II in German historiography. Frederick is so central to the modern history of Prussia that the cataclysms of 1914–45 could not but affect deeply his historical standing. Catherine, a foreigner by birth and in less direct personal contact with the problems of her vast empire (there is no counterpart in her case to the regular journeys of inspection in the Prussian provinces to which Frederick devoted so much time and energy) seems to stand in many ways apart from the main stream of Russian life as neither the king of Prussia nor Joseph II did from that of their territories.

In her own day the empress enjoyed in general a good press and a high reputation. This was in part her own doing. To a considerable extent it was the result of a deliberate public relations campaign, at least in France, then the leader of European taste and opinion in so many ways. Catherine's efforts to present herself to the outside world as a supporter of progressive ideas in politics and as a patron of scholars and artists had the powerful backing of intellectuals so influential as Diderot and above all Voltaire.[1] They were also favoured by the implied compliment of ruling-class Russia's ostentatious imitation of western Europe; by a general taste, at least in France, for the exotic and the big; by the prestige with which military and diplomatic success inevitably endowed any ruler; by a growing

---

Russkaya istoriografiya (Moscow, 1948); M. V. Nechkina and others (eds.), *Ocherki istorii istoricheskoi nauki v SSSR* (4 vols., Moscow, 1955–66); S. L. Peshtich, *Russkaya istoriografiya XVIII veka* (2 vols., Leningrad, 1961–5). The best guide to the work published in the 1960s is probably M. Raeff, 'Random Notes on the Reign of Catherine II in the Light of Recent Literature', *Jahrbücher für Geschichte Osteuropas*, Neue Folge, Band 20 (1972), No. 1, 541–55.

[1] See on this the now classic work of Lortholary, *Le Mirage russe*, pp. 150–70. A well-known instance of the empress's sensitivity to unfavourable publicity is her repeated efforts to suppress the manuscript of C. C. de Rulhière's *Histoire, ou anecdotes sur la révolution de Russie en l'année 1762*. Though the author successfully resisted both threats and a bribe of 30,000 francs to delete certain parts of what he had written he was compelled to agree that the book should not appear during Catherine's lifetime. It was not in fact published until 1797. See the 'Note de l'éditeur', pp. ii–iii.

feeling among the *philosophes* that the north of Europe was now more advanced and enlightened than the south;[1] and perhaps most of all by the contrast which Catherine's energy and enterprise seemed to offer to the lethargy and ineffectiveness of Louis XV and Louis XVI.[2] Even in the first decade of her reign publicists had begun to picture her as the heir to the policies and aspirations of Peter the Great and as continuing what was seen as his great work of civilizing an immense semi-barbaric empire and bringing it into the mainstream of civilization.[3] This flattering view of Catherine as both progressive in her politics and glorious in their execution, both enlightened and Roman, remained influential throughout her reign and for long afterwards.

Nevertheless there were other aspects of her achievements which, in the eyes of contemporaries, shadowed the brilliance of her success. What to most writers of the last hundred years has seemed much the blackest of these shadows, the intensification and geographical extension of serfdom, attracted little attention. The meaning of this development for the Russian people was simply not understood at all. Criticisms of Catherine tended to concentrate much more on specific personal issues—the way in which she had gained power in 1762 by the overthrow and murder of her husband; her personal morals; the corruption, particularly in her later years, of some of her ministers and favourites. These were enough, however, to ensure that many contemporary judgements of her were balanced and even ambiguous, though relatively few were downright unfavourable. Among many denunciations of her treatment of her husband and conventional criticisms of her licentiousness and extravagance[4] a few writers in western Europe were able to probe rather deeper. Some grasped the extent to which vanity and the desire of reputation were a mainspring of her actions. One observer remarked tellingly in the first years of her reign that 'there is a sort of whim or affectation of singularity, in the manner of conferring her favours, that looks as if the desire of being spoken of, fully as much as the desire of doing good,

[1] On this see Lortholary, *Le Mirage russe*, pp. 167–70.

[2] On the workings of these factors see the somewhat acid comments of C. C. de Rulhière, *Histoire de l'anarchie de Pologne, et du démembrement de cette royaume* (2nd edn., Paris, 1808), iii. 142.

[3] e.g. von Schlözer, *Neuverändertes Russland oder Leben Catherina der Zweyten*, i, *Vorrede*.

[4] e.g. P.-C. Levesque, *Histoire de Russie et des principales nations de l'empire russe*, (4th edn., Paris, 1812), vi. 69–70; W. Anderson, *Sketches of the History and Present State of the Russian Empire* (London, 1815), pp. 370–1.

was the fountain from which they flow'.[1] Others realized that the great political and military successes of Russia under Catherine's rule were not the result of deep planning on her part but of a combination of skill, self-confidence, and mere luck which had allowed her to overcome serious difficulties.[2] Others again showed some understanding of the vulnerability of her position in Russia and of the limits which her position as a usurper and a woman placed on her real authority. 'How', asked a somewhat scandal-mongering but well-informed Frenchman, 'was a woman to effect that which the active discipline of the cane, and the sanguinary axe of Peter I, were inadequate to accomplish?' After all 'it was solely by suffering her power to be abused, that she succeeded in preserving it'.[3] It cannot be claimed, however, that contemporary or near-contemporary west European comment on Catherine shows any great penetration. Too many sides of Russian life and too many of the implications of her reign went still unperceived.

Russian evaluations of Catherine during this period, in a society still dominated by an oppressive autocracy and with little tradition of historical analysis, were few in number. They are however at least equal in depth to anything produced in western Europe and far from indiscriminate in their admiration for her. The extreme conservative Prince M. M. Shcherbatov, in his *O povrezhdenie nravov v Rossii (On the Corruption of Morals in Russia)*, written in 1786–7, launched against Catherine from the standpoint of the old Russian nobility one of the most severe criticisms made of her during her lifetime. He agreed that she had many good qualities; but the puritanism and religious conservatism which largely dominated him found much to deplore in her. 'Her moral outlook', he wrote, 'is based on modern philosophers, that is to say, it is not fixed on the firm rock of God's Law; and hence, being based on arbitrary worldly principles, it is liable to change with them. . . . Carried away by her indiscriminate reading of modern writers, she thinks nothing of the Christian religion, though she pretends to be quite devout.'[4] As

---

[1] W. Richardson, *Anecdotes of the Russian Empire, in a Series of Letters written a few Years ago, from St. Petersburg* (London, 1784), p. 26. This letter was written in November 1768.

[2] Rulhière, *Histoire de l'anarchie de Pologne*, i. 376–8.

[3] *Secret Memoirs of the Court of St. Petersburg* (2nd edn., London, 1801), p. 61.

[4] Prince M. M. Shcherbatov, *On the Corruption of Morals in Russia*, translated and edited by A. Lentin (Cambridge, 1969), pp. 235, 257.

a result, she was self-indulgent and capricious, fond of flattery, and sexually immoral. More serious from a political standpoint, she was dominated by a thirst for glory and reputation (the one characteristic which almost every contemporary observer agreed in attributing to her). This was the real explanation of her intervention in Poland and her annexation of the Crimea.[1] The deeply felt character of these criticisms is unmistakable. It was the product in part of Shcherbatov's failure to achieve the success and official rank to which he felt his talents entitled him. It also sprang from his membership of a very ancient noble family and resulting dislike of the new nobility, based not on ancestry but on official rank and state service, which had rapidly become more powerful in Russia since the reign of Peter I. In him political conservatism, which aimed at expelling these upstarts from the nobility and at limiting the power of the autocrat and strengthening the Senate and the old noble families, was coupled with a puritanical dislike of 'luxury' (here he represents an important strain in the 'enlightened' psychology of the later eighteenth century) and a deep hostility to foreign influences in Russia. The combination could not but inspire hostility to many aspects of Catherine's life and regime.

Shcherbatov, however, for all his abilities, was too individual and too opinionated to be representative of Russian ruling-class feeling. This was more truly reflected in the comments on Catherine in the *Zapiska o drevnei i novoi Rossii* (*Memoir on Ancient and Modern Russia*) of the official historian N. M. Karamzin. (It was written in early 1811 but was not intended for publication and was not published in full until 1861.) Karamzin was a conservative of a much more conventional type than Shcherbatov. As a member of a gentry family of the kind which had benefited from the whole drift of Russian history over the last century, and notably from some of the legislation of Catherine's reign, he was in essence an upholder of the *status quo* and correspondingly more generous in his assessment of the empress's achievements. He did not conceal his distaste for her personal life, which made him 'blush for mankind'; and he repeated the criticism made more than once by foreign observers, that her innovations were often designed for appearance and reputation rather than real effect. 'The very political institutions devised by Catherine', he wrote, 'reveal more sparkle than substance; the choice

---

[1] **Shcherbatov**, *Corruption of Morals*, pp. 241–55, *passim*.

fell not upon the best in content, but the prettiest in form. . . . We possessed academies, institutions of higher learning, popular schools, wise ministers, heroes, a superb army, an illustrious fleet, and a great queen—but we lacked decent upbringing, firm principles, and social morality.'[1] This was a formidable indictment. Yet Karamzin's verdict was on balance strongly favourable. Catherine had, after all, been the heir of Peter the Great, 'the true inheritor of Petrine greatness and the second architect of the new Russia'. She had 'cleansed autocracy of the stains of tyranny'; and 'when all is said and done, should we compare all the known epochs of Russian history, virtually all would agree that Catherine's epoch was the happiest for Russian citizens; virtually all would have preferred to have lived then than at any other time'.[2]

The modern reader must be struck by what seem to him the superficiality and the class bias of both Shcherbatov and Karamzin. Different in so many ways, they are completely at one in their indifference to the needs and sufferings of the great mass of the Russian people. To them Russia means the court, the higher ranks of the administration, the armed forces, the church, and a ruling class of landowners (however much they differ about the proper bounds to be set to this class). The peasant and even the ordinary townsman scarcely appear at all in their scheme of things. Not only are these groups excluded from the 'political nation'; they are not really part of Russian history. This class-bound and, in Karamzin's case, officially-biased view betrays itself most obviously to modern eyes in the fact that neither has anything at all to say about what today seems the most important of all the results of Catherine's reign for the internal development of Russia—the extension and intensification of serfdom which it witnessed. This restricted and official picture of the period, this view of Russian society solely from above, was to influence a considerable amount of historical writing throughout the nineteenth century and down to the 1917 revolution. The two volumes (of a never-completed twelve-volume biography) which in the 1890s V. A. Bil'basov devoted to Catherine's first years as ruler are a good case in point. Full of anecdotes, of information about court intrigues and ruling-class rivalries, they contain much interesting material but present a narrow view not only of Russian life but of Catherine's

[1] R. Pipes (ed.), *Karamzin's Memoir on Ancient and Modern Russia* (Cambridge, Mass., 1959), p. 133.
[2] *Karamzin's Memoir on Ancient and Modern Russia*, pp. 130, 131, 134.

position and problems.[1] At the level of the school textbook and the
student's manual this narrowness of view was particularly marked.
The third quarter of the nineteenth century, when the lower and
middle reaches of the Russian educational system were expanding
rapidly, saw a large output of works of this kind which stressed one-
sidedly the importance of Russian rulers and their activities and the
constructive significance of the autocracy; many of these went
through numerous editions.[2] However there was no shortage of
favourable judgements of Catherine even from scholars of the highest
calibre. Thus V. O. Klyuchevskii, the greatest historian of Russia,
in the last volume of his *Kurs russkoi istorii*, which was still incomplete
at his death in 1911, devoted much attention to the increasing scope
and weight of serfdom under her rule. Yet this did not prevent him,
in the essay on her which he wrote in 1896, from taking a very high
view of her achievements. Whatever her shortcomings, her power
had been used to achieve external safety and internal peace for
Russia, and she had been 'the advocate of liberty and enlightenment'.[3]
In rather the same way the liberalism (real or ostensible) and the
genuine cultural interests of Catherine, which had gained her admiration from the first days of her reign, won the approval of S. F.
Platonov, one of the ablest of twentieth-century Russian historians.
He saw her as 'schooled in the liberating theories of the eighteenth
century' and claimed that only the opposition of conservative ministers
had compelled her to abandon the dream of freeing the peasants from
the yoke of serfdom. She had been 'a historical figure of first-class
importance' who had aimed at 'the complete solution of the problems
with which history faced her'.[4]

By the second half of the nineteenth century, however, concern with
the peasant problem, a concern aroused rather than assuaged by the
emancipation edict of 1861, was becoming more and more the key-
note of a great deal of Russian intellectual life. The growth of a
populist movement devoted to the improvement of the lot of the
peasantry, on whose labours the growing intelligentsia class was
more and more conscious of its dependence, was the most striking
feature of Russian political life in the 1860s and 1870s. This naïve
but sincere idealization of the peasant was seen in its most striking

[1] V. A. Bil'basov, *Istoriya Ekateriny Vtoroi* (2 vols., 1890–1).

[2] See the details given in Nechkina, *Ocherki istorii*, ii. 102.

[3] *Kurs russkoi istorii* (Moscow, 1937), v. 164–5; Nechkina, *Ocherki istorii*, ii.
157.

[4] *Lektsii po russkoi istorii*, (9th edn., Petrograd, 1915), pp. 621–2.

form in the 'going to the people' of the early 1870s, when thousands of students and other young enthusiasts from the cities attempted vainly, by living in the villages, to educate and indoctrinate rural society. The same idealization also had a profound effect on the writing of Russian history. So far as the reign of Catherine II was concerned the most important single product of this new influence was the remarkable book of V. I. Semevskii, *Krest'yane v tsarstvovanie imperatritsy Ekateriny II* (*The Peasantry during the Reign of the Empress Catherine II*) which was published in St. Petersburg in 1881. It was followed seven years later by an even more exhaustive work, *Krest'yanskii vopros v Rossii v XVIII i pervoi polovine XIX v.* (*The Peasant Question in Russia during the Eighteenth and first half of the Nineteenth Century*). Semevskii was by no means the only populist historian to devote his energies to the peasantry; but his work alone has remained important to the present day.[1] No scholar has been inspired by a higher or more urgent sense of moral obligation. To him the writing of his books was the payment of a sacred debt which he, in common with all other Russian intellectuals, owed to the largest, the most despised, and yet the most essential group in society. 'Is it not time to give serious attention to the history of the people, and not merely to that of the court or of diplomacy?' he wrote in a famous article in 1881. 'The history of the peasantry in general, and in our country in particular, ought to take first place, not merely because this class is so numerous but because of the significance which it has for the state'. The writing of this history was an inescapable obligation for the intelligentsia which 'is bound to work for the benefit of the peasant both in everyday life and in scholarship'; for 'the history of the Russian people is the debt owed by our scholarship to the people'.[2] Study of the history of the peasantry was even more than an obligation; it was also to Semevskii an essential weapon in the struggle to improve the position of the peasant. In his speech in defence of his master's thesis at Moscow University in 1882 he proclaimed his belief that 'the present can be properly understood only in the searching light of history'.[3]

[1] There is a useful short study of his life and ideas in M. B. Petrovich, 'V. I. Semevskii (1848–1916): Russian Social Historian', in *Essays in Russian and Soviet History in Honor of G. T. Robinson*, ed. J. S. Curtiss (Leyden, 1965), pp. 63–84. For a Soviet view see Nechkina, *Ocherki istorii*, ii. 207–18.

[2] 'Ne pora-li napisat' istoriyu krestyan' v Rossii?' *Russkaya Mysl'*, 1881, No. 2, 215–65.

[3] *Russkaya Starina*, May 1882, p. 577.

Beliefs such as these, openly declared, were bound to arouse uneasiness in the government of the day. In 1886 Semevskii was dismissed from his post at the University of St. Petersburg and never taught formally again. Nevertheless his achievement was great. Far more than any other historian he made the peasant question, hitherto astonishingly neglected, the central problem of Russian historiography. Since the 1880s there has been a vast output of scholarly writing on the history of the Russian peasantry, much of it of high quality, under both the imperial and Soviet regimes. This stems above all from Semevskii's work and example. Before him studies of the subject had been limited in their views, in their choice of material, and in their geographical scope. They had usually confined themselves to the analysis of official enactments regarding the status and obligations of the peasant, or to particular areas of Russia or specific categories of peasant. Semevskii assembled an enormous amount of new material and attacked the subject on a far broader front than his predecessors. His comments on Catherine were not in general hostile. Indeed he took a favourable view of her as the initiator of the struggle for peasant freedom. 'It is impossible', he wrote, 'to give her higher praise than that she clearly posed the question of the inevitability of an improvement of the position of the peasants on private estates.' By acting in this way 'she . . . inscribed her name for ever on the history of the peasant question in Russia'.[1] But he enunciated clearly for the first time a charge against the empress which was to be repeated and elaborated down to the present day: the accusation that her reign had seen serfdom become more onerous and more widespread in Russia than ever before. Increasingly this was to become, particularly in the writings of Soviet historians, the most serious of all indictments of her, the great blot on her reign.

As well as this serious accusation, the nineteenth century also saw foreshadowings of another which has been made more and more frequently in recent decades—the claim that Catherine's liberalism and enlightenment were a mere sham adopted to present an attractive façade to the gullible west. This idea was implicit to some extent in the feeling visible in her own lifetime that Catherine's reforms were inspired by 'the desire of being spoken of, fully as much as the desire of doing good'.[2] In our own day argument on this point

---

[1] *Krest'yanskii vopros*, i. xiv, 228.
[2] See above, pp. 160-1.

has raged freely though inconclusively.[1] During much of the nine-teenth century the question of the sincerity of Catherine's enlighten-ment was hardly in the forefront of discussion; but even then it aroused attention. The most famous of all Russian liberals, Alexander Herzen, had no doubt that power, not progress or welfare, was the driving-force which inspired Catherine. She had 'sacrificed every-thing, and above all the happiness of the people, to create a Russian empire, to organize a powerful government and make it European'. In the last years of her reign, terrified by events in France, 'she frankly tore off all appearance of liberalism . . . and revealed herself finally for what she was in reality: an old heartless Messalina, a Lucrezia Borgia with German blood in her veins'.[2]

To a considerable extent, however, discussion of Catherine's personal influence on the events of her reign was by the end of the nineteenth century being pushed into the background by studies of special problems or aspects of the period. These were now of a kind more detailed and professional than any yet seen in Russia. In this respect the two decades before the Revolution of 1917 were a golden age of Russian historiography, a period during which it reached a level quite comparable to anything attained in France or Germany and higher than that achieved in Britain. A. V. Florovskii, for example, made the first important contributions to the study of a subject still unexhausted; the significance of the Legislative Com-mission of 1767, the most spectacular of all Catherine's liberal initia-tives.[3] Almost simultaneously A. A. Kizevetter produced a detailed study of the emergence of the great decree of 1785 which had reorganized town government in Russia,[4] while a few years earlier N. N. Firsov had produced an important discussion based on ex-tensive archive work of Russia's foreign trade and the governmental and social background to it.[5] This list could easily be lengthened; but the point need not be laboured. In Russia as elsewhere the professional academic, scrutinizing archives to produce detailed studies of relatively small and manageable subjects, had now come

---

[1] See below, pp. 169–71.

[2] A. I. Herzen, *Sochineniya*, ix. 206–7.

[3] *Iz istorii ekaterinskoi zakonodatel'noi komissii 1767g. Vopros o krepostnom prave* (Odessa, 1910); *Sostav zakonodatel'noi komissii 1767–74gg.* (Odessa, 1915).

[4] *Gorodskoe polozhenie Ekateriny II 1785g. Opyt istoricheskogo kommentariya* (Moscow, 1909).

[5] N. N. Firsov, *Pravitel'stvo i obshchestvo v ikh otnoshenii k vneshnei torgovlye Rossii v tsarstvovanie imperatritsy Ekateriny II* (Kazan, 1902).

to dominate completely the writing of serious history. In Russia as elsewhere this development was associated with the systematic publication of documents on a hitherto unknown scale. In particular the Imperial Historical Society, in the 148 volumes of its *Collection* (*Sbornik*) which appeared between 1867 and 1916, made available to historians a very large amount of material relating to Catherine's reign.[1] This was markedly governmental in character and related overwhelmingly to official issues and problems, such as the work of the Legislative Commission or government finance. Much of it was also incomplete. The publication of Catherine's personal correspondence by the Imperial Historical Society, interrupted by the Revolution, remained very fragmentary; and although much of her considerable literary output appeared between 1901 and 1907 in a scholarly edition under the editorship of A. N. Pypin, this enterprise also was never completed. Nevertheless in Russia, at least as much as in France or Germany, the new temper of historical writing showed itself in the voluminous collection of printed documents as well as in the scholarly monograph.

To the study of Catherine and her reign the contributions made by the Soviet historical writing of the last two generations is in general disappointing. It has produced little comparable in scope and quality to the best work of the generation before 1917; nor has there been any sustained publication of documents of the kind undertaken by the Imperial Historical Society. There has, perhaps inevitably, been a continuing and ever-intensified interest in the peasantry, in the burdens placed upon them and their reaction to these burdens.[2] One episode in particular of peasant struggle, the most violent and spectacular of all, has attracted attention. This is the great revolt led by Emelyan Pugachev in 1773–5 which, largely because of the ease with which it fits into a system of historical explanation based on the idea of class conflict, has generated a large body of writing; though much of this has been weakened by a marked tendency to exaggerate the potentialities of 'revolutionary' forces in Russia under Catherine's regime.[3] Of more far-reaching importance, there has during the last

---

[1] *Sbornik Imperatorskogo Russkogo Istoricheskogo Obshchestva* (St. Petersburg, 1867–1916).

[2] The most complete Soviet study of this aspect of the subject is N. L. Rubinshtein, *Sel'skoe khozyaistvo Rossii vo vtoroi polovine XVIIIv.: istoriko-ekonomicheskii ocherk* (Moscow, 1957).

[3] J. T. Alexander, 'Recent Soviet Historiography on the Pugachev Revolt: A Review Article', *Canadian Slavic Studies*, iv. No. 3 (Fall, 1970), 602–71, provides

decade been considerable discussion, which has obvious relevance to the reign of Catherine II, of the nature of absolutism in Russia. This has revealed an interesting readiness to contemplate some movement away from a rigid or traditional Marxist explanation of the pheno-menon.[1] The orthodox view, tracing mainly from Engels, which sees absolute monarchy as a transitional form of government resulting from a temporary balance in society between declining feudal and rising capitalist forces, has been increasingly called in question by a number of Soviet historians, notably N. I. Pavlenko and A. Ya. Avrech. The unmistakable weakness of capitalism in eighteenth-century Russia has made it difficult to visualize absolutism there as the product merely of the growth of a capitalist mode of production. Instead it can now be seen as resulting from the rise of the state-structure to a level higher than that achieved by the underlying economy and society, or alternatively as the result of an imperative need to defend Russia against pressure from the more advanced western powers. The debate has been confused and, up to the moment, inconclusive. But it is significant as showing at least the beginnings of an understanding of the limitations of a strictly Marxist analysis of the question. Yet the Soviet historiography of this period has been, with few exceptions, limited in scope and subject and often stereotyped and over-schematic.

So far as Catherine herself is concerned the main charge levied against her by Soviet historians is the already well-worn one of hypocrisy. More insistently than their predecessors they have stressed the allegedly glaring contrast between the liberal and enlightened ideals which she proclaimed and the oppressive despotism which she practised. Her regime, it is argued, was in essence an expression and defence of the class-interests of the landowning ruling class. The discontent of the great suppressed and exploited majority, the urban bourgeoisie and above all the peasantry, was concealed and camouflaged, within Russia and even more abroad, by a false but

---

full references. The most important Soviet work on Pugachev and his followers is V. V. Mavrodin (ed.), *Krest'yanskaya voina v Rossii v 1773–1775 gg.* (3 vols., Moscow, 1961–70).

[1] See A. Gerschenkron, 'Soviet Marxism and Absolutism', *Slavic Review*, December 1971, pp. 853–69: the most significant contribution to this controversy has probably been A. Ya. Avrech, 'Russkii absolyutizm i ego rol' v utverzhdenii kapitalizma v Rossii', *Istoriya SSSR* (1968), No. 2, 82–104. The collection of essays edited by N. M. Druzhinin, *Absolyutizm v Rossii* (Moscow, 1964), which contains much useful information, appeared before theoretical discussion of the subject became really active.

ostentatious concern for progress and enlightenment.[1] West European writing, on the other hand, though far from uncritical of Catherine, has been more ready to take her professions at something like face value. Many western historians have thought in terms of at least a few years of genuine liberalism at the beginning of her reign, even if from the end of the 1760s this impulse towards enlightened reform was weakened and stultified by the Turkish war and the Pugachev revolt, and finally extinguished by the French Revolution.[2] It has been argued in an important recent study that the empress's genuine attachment to ideals of individual liberties and intellectual freedom remained constant until the last years of her reign, and that even at her death, in spite of the bitter conservative reaction provoked by the French Revolution, there was much more freedom of expression in Russia than in the 1750s.[3] The non-Soviet approach to Catherine's motives seems obviously more realistic, more in accordance with the facts of ordinary human experience, than the Soviet one. Every statesman must be guilty of a certain minimum of necessary hypocrisy; this is the unavoidable price he pays for his position. But to imagine that Catherine spent a lifetime in unremitting and explicit hypocrisy, with all the psychological strain this must involve, is as unfair to her intellectual integrity as it is unduly favourable to her strength of character. The safest and fairest conclusion is that she had sincere impulses towards reform which the difficulties and vulnerability of her position, and a highly-developed sense of self-preservation, prevented her from realizing more than very partially, even in the optimistic early years of her reign.

Historical writing in the non-Soviet world, like that in Russia, has been unable to escape from a concern with the plight of the peasant, and above all the serf, under Catherine's rule. But the attitudes to the peasant question taken by western historians have been much more varied and indeed more confused than those adopted by Soviet writers. On the one hand there is agreement that Catherine did nothing effective to reduce the burdens which serfdom imposed on the

[1] For a useful list of recent Soviet works which have adopted, in varying ways and to varying extents, this line of argument, see D. M. Griffiths, 'Catherine II: The Republican Empress', *Jahrbücher für Geschichte Osteuropas*, Neue Folge, Band 21 (1973), 323–4, fn. 4.

[2] For a list of books and articles which argue in this sense see Griffiths, 'Catherine II', 324, fn. 5.

[3] K. A. Papmehl, *Freedom of Expression in Eighteenth Century Russia* (The Hague, 1971), p. 132.

Russian peasant, and a great deal to widen its geographical scope, notably by the decree of May 1783 which completed its extension to the Ukraine. On the other hand it had already been well established in Russia before the Revolution that she clearly made considerable efforts to prevent its illegitimate or unintended growth, by preventing the adscription to landowners during the process of census-taking of people hitherto free, such as orphans or unassigned clergy, and by limiting the enserfment of war prisoners or enserfment by marriage.[1] This and other pleas in mitigation of Catherine's alleged guilt for the damage inflicted by serfdom on Russian society have recently been repeated with considerable force by a number of writers in western Europe and the United States. It has been argued, for example, that the halting of peasant movement in the Ukraine and the extension there of the poll-tax, which doomed the Ukrainian peasant to serf status, were inspired by a desire to end the privileges enjoyed by many border areas of the empire, and to a lesser extent by the need for increased revenue, rather than by the extension of serfdom as a matter of policy. In the same way the decree of 1767 which forbade the submission of petitions directly to the empress has been shown to be a mere repetition of earlier legislation and in any case not specifically directed against peasant petitioners.[2] The most complete and recent defence of Catherine's attitude to serfdom concentrates on showing convincingly that much of her legislation on the question was complex and confused and its effects hard to measure.[3]

There is at least one obvious reason why the defects of Catherine's domestic policies may be expected sometimes to weigh rather less heavily with a west European or American writer than with most Russian ones. The former, seeing Russia from the outside, can scarcely fail to be aware of the brilliant success of the empress's foreign policy and of the military achievements and territorial growth of Russia under her rule. A very high proportion of Russian historians from the 1860s or 1870s onwards, however, have been unwilling or unable to balance Catherine's achievements in these spheres against her weaknesses and limitations in others. Neither populists like

---

[1] A. S. Lappo-Danilevskii, 'Ekaterina II i krest'yanskii vopros', in *Velikaya reforma* (Moscow, 1911), translated in M. Raeff (ed.), *Catherine the Great: A Profile* (London, 1972), 267–89.

[2] Griffiths, 'Catherine II', pp. 329–30.

[3] Isabel de Madariaga, 'Catherine II and the Serfs: A Reconsideration of some Problems', *Slavonic and East European Review*, lii, No. 127 (1974), 34–62.

Semevskii nor liberals such as Kizevetter[1] were at all interested in diplomatic or military history; and their concentration on domestic, social, and economic questions, together with their lack of understanding of the realities of power, of the limitations which the need to defend Russia and maintain some effective central government placed upon Catherine's freedom of action, have been inherited by a high proportion of their Soviet successors. More fundamentally, it is unfair and unhistorical to expect Catherine, in an age when Russian and most of European society was organized in terms of distinct social estates with differing legal rights and obligations, to act like a nineteenth-century west European liberal. Her ideal, it has been suggested in two important articles by a German historian, was one strictly of her own age—that of building up and strengthening the nobility, the still largely embryonic middle class, and to some extent the state peasants (though not the privately-owned serfs) as efficient and creative autonomous groups. These she hoped would prove capable of rescuing Russian society from intellectual backwardness and agrarian routine. The provincial reforms of 1775 and the great grants of privileges to the nobility and towns in 1785 were essentially means to this end.[2] This line of thought, whatever the status eventually assigned to it as the historiography of Catherine II develops, has at least the merit of directing attention very clearly to the positive intentions of her complex legislation and away from the more negative issue of her failure to improve the status of a particular, though extremely large and important, group in Russian society.

To the majority of historians outside Russia, however, the workings of the machinery of government under Catherine and the empress's efforts to improve them have seemed during the last half-century a more interesting subject than the details of peasant suffering. The Legislative Commission of 1767 and the *Nakaz* which laid down the

---

[1] A. A. Kizevetter, 'Ekaterina II', in *Istoricheskie siluety* (Berlin, 1931), translated in M. Raeff (ed.), *Catherine the Great*, pp. 3–20. This is the most complete statement by any major twentieth-century historian of the argument that Catherine was dominated by a desire for reputation, especially outside Russia, and a drive towards self-advertisement. For a rather similar view by another great Russian émigré historian see P. N. Milyukov, 'Catherine II', in A. B. Duff and F. Galy (eds.), *Hommes d'état*, iii (Paris, 1936), 22–3.

[2] D. Geyer, 'Staatsaufbau und Sozialverfassung—Probleme des russischen Absolutismus am Ende des 18. Jahrhunderts', *Cahiers du monde russe et soviétique*, vii. (1966), 366–77; and 'Gesellschaft als staatliche Veranstaltung (Bemerkungen zur Sozialgeschichte der russischen Staatsverwaltung im 18. Jahrhundert)', *Jahrbücher für Geschichte Osteuropas*, xiv (1966), 21–50.

general guide-lines for its work have continued to attract much, and on the whole favourable, attention. Few historians would now agree with the claim of W. F. Reddaway, made more than four decades ago, that Catherine 'brought Russia nearer to freedom in 1767 than did Alexander I after overthrowing Napoleon, or Alexander II after the defeat of Russian absolutism by the West, or Nicholas II after the unprecedented humiliations of 1905'.[1] Yet a more recent and far better-informed writer has claimed that the *Nakaz* was 'a sophisticated rationale for monarchy without despotism' and that if the impetus towards reform which Catherine gave the mechanism of government had been maintained by her successors 'a nineteenth-century *Rechtstaat* on the Prussian pattern might have emerged'.[2] Another recent student has concluded that the attempted reforms of the empress's first years were in reality a largely successful effort to create a rationalized bureaucracy capable of governing Russia without, as in the past, calling on the army for assistance. 'In their effect,' he concludes, 'the reforms conducted in the early years of Catherine II's reign amounted to an administrative revolution. They created a civil administration such as had not previously existed in eighteenth-century Russia.'[3]

The same note of marked and even enthusiastic approval can be heard in most of the increasingly expert comment of recent years on the administration of the empire. Catherine's consistent efforts to improve the workings of government in the provinces, particularly after the Pugachev revolt, have been stressed; though it is clear that the mere smallness of the bureaucracy in proportion to the enormous size of the country was here a fundamental difficulty.[4] The creation during her reign of a relatively efficient centralized financial system in place of the cumbersome decentralized collegiate one inherited largely unchanged from the days of Peter I has drawn praise: by 1781, the most recent writer on the subject has claimed, it was possible to submit to her 'the first meaningful budget in the history of Imperial Russia'.[5] The government-directed planning under Catherine of over

---

[1] *Documents of Catherine the Great* (Cambridge, 1931), Introduction, p. xxvi.

[2] Griffiths, 'Catherine II', 325, 327.

[3] K. R. Morrison, 'Catherine II's Legislative Commission: An Administrative Interpretation', *Canadian Slavic Studies*, iv (1970), 472–3.

[4] R. E. Jones, 'Catherine II and the Provincial Reform of 1775: A Question of Motivation', *Canadian Slavic Studies*, iv (1970), 497–512.

[5] J. A. Duran, Jr., 'The Reform of Financial Administration in Russia during the Reign of Catherine II', *Canadian Slavic Studies*, iv (1970), 493.

four hundred new cities and towns has been put forward, in an interesting and original article, as 'one of the important legacies of the eighteenth century in Russia'.[1] And these are merely a very arbitrary and random selection of articles published in the last few years in one major western language. The good press abroad which Catherine so much desired she has now very largely attained. It is based, moreover, on a knowledge of her achievements and under-standing of her problems far more detailed and complete than was ever possible in her own lifetime.

Two more developments in the way in which historians have now begun to see Catherine II and her Russia deserve brief mention. The first of these is a growing emphasis on the extent to which her policies and methods were inspired by purely Russian necessities, the way in which they were rooted in the work of her predecessors and not in theories imported from western Europe. This view of her work, which sharply contradicts the belief, so widespread in her reign and to a large extent encouraged by the empress herself, that she aimed above all at the application in Russia of ideas drawn from foreign sources, had already been put forward before the 1917 Revolution. But the whole drift of twentieth-century writing has tended to lay increasingly heavy stress on it and to see Catherine, rightly, more and more in terms of Russian conditions rather than in those of the Enlightenment as a general European intellectual phenomenon. Nearly half a century ago, in a famous essay, Kizevetter emphasized that even the reforming programme of her early years, when foreign influences on her are usually considered to have been at their strongest, was a Russian one in which these influences played little part. In Catherine's policies, he alleged, 'the trappings of philosophical ideology' were a mere surface adornment. Not all students would go so far as this. But few would now seriously challenge the claim that when action was concerned it was to the precedents of the reign of Elizabeth (notably the creation then of a commission for the drawing-up of a new law code) rather than to foreign ideas of any kind, that Catherine turned.[2]

Perhaps the most interesting and promising of all recent trends in the study of the Russia of Catherine the Great, however, has been the effort made by one or two writers to describe and analyse in

[1] R. E. Jones, 'Urban Planning and the Development of Provincial Towns in Russia, 1762–1796', in J. G. Garrard (ed.), *The Eighteenth Century in Russia* (London, 1973), p. 321.

[2] A. A. Kizevetter, 'Pervoe pyatiletie pravleniya Ekateriny II', in *Istoricheskie siluety* (Berlin, 1931), translated in M. Raeff (ed.), *Catherine the Great*, pp. 247–66.

detail the psychology of different social groups and the way in which this determined their behaviour, especially in economic life. This study of group psychologies and life-styles parallels to some extent, on a smaller scale, some of the impressive work which has already been referred to in the case of France.[1] In two remarkable books, *Domaines et seigneurs en Russie vers la fin du XVIIIe siècle. Étude de structures agraires et de mentalités économiques* (Paris, 1963) and *Systèmes agraires et progrès agricole. L'Assolement triennial en Russie aux XVIIIe-XIXe siècles* (Paris–The Hague, 1969), Michael Confino has shown with unprecedented clarity and completeness how far the Russian landowner of this period was from being a rational 'economic man' and how greatly his behaviour was influenced by tradition, habit, ignorance, and non-economic forces in general. Moreover, as the second of these books in particular makes clear, even the small minority of forward-looking landowners interested in agricultural progress encountered almost insuperable obstacles from peasant conservatism and resistance to change. The Russian village community, in other words, cannot be adequately described simply in the rationalizing Marxist terms of conflicting economic interests. With all its gross injustices and brutalities it was a balanced (though usually precariously balanced) whole, in which change of any kind was a threat to the balance and correctly seen as such. It was often the cultural and psychological gap which yawned between the peasant and a progressive landlord wishing to exploit his estates and his labour-rights more effectively, as much as any material exploitation, which inspired agrarian revolt. The subtle and many-dimensioned view of their subject offered by these books represents a real advance in the study of Russian society in the age of Catherine the Great. A similar sublety and imagination, on a smaller scale, can be seen in other recent publications. W. R. Augustine, in a remarkable study of the economic attitudes of the nobility of central Russia in the 1760s[2] has underlined their relative lack of interest in material gain for its own sake, the extent to which the attainment of a satisfactory rank in state service and the maintenance of a life-style in keeping with that rank took precedence over any idea of the strictly rational exploitation of their estates. The noble's position as a landowner was the necessary foundation of a satisfactory way of life rather than a

---

[1] See above, pp. 52–8.
[2] W. R. Augustine, 'Notes toward a Portrait of the Eighteenth-Century Russian Nobility', *Canadian Slavic Studies*, iv, No. 3 (Fall 1970), 373–425.

means of becoming rich (an attitude which does more than anything to explain the chronic indebtedness which was henceforth to undermine and eventually destroy the position of the Russian landowning class). In a somewhat different way Professor Raeff has made a similar point, by showing how state service, which remained after 1762 a moral and social obligation for the nobility even though it had ceased to be legally compulsory, influenced the whole outlook of the landowning ruling class. It endowed that class as the century progressed with a growing consciousness of the extent to which the Russian state depended upon it and an increasingly full realization of its own value and importance.[1] He also raises a still largely unstudied question of great importance—that of the differences of outlook which undoubtedly existed on some issues between the members of the ruling class in different areas of Russia. The landowners of the Volga and Ural area, who had a strong sense of taking part in a centuries-old forward movement of Russian power and civilization; those of south Russia, proud of the victories their country had achieved over the Turks and Tatars; those of western Russia, often strongly anti-Polish; those of the German-dominated Baltic provinces: these groups, at least in their attitudes to the world outside Russia, differed as much as they agreed. The study of regional differences of this and other kinds in the Russia of Catherine II is still in its infancy: it is one of the most promising avenues of advance for the future.

THE HABSBURG EMPIRE: MARIA THERESA AND JOSEPH II

In the historiography of eighteenth-century Europe the Emperor Joseph II, the third of the trio of 'enlightened despots' to have achieved canonization in even the most elementary of present-day textbooks, occupies a position in some respects midway between those of Frederick II and Catherine II. The fact that he was sole ruler of the Habsburg lands for only the decade 1780–90, and more important the failure in practice of many of his policies, have prevented his achieving quite the symbolic importance, the power to generate both passionate loyalty and deep dislike, attained by the Prussian king. On the other hand he still stands out unmistakably, as he did

---

[1] M. Raeff, 'Staatsdienst, Aussenpolitik, Ideologie (Die Rolle der Institutionen in der geistigen Entwicklung des Russischen Adels im 18 Jahrhundert', *Jahrbücher für Geschichte Osteuropas*, Neue Folge, vii (1959), 147–81.

in his own lifetime, as a distinctive individual, more clearly so than almost any other major ruler of the century. Unlike Catherine II, he has never been pushed completely into the background by historians interested only in the social and economic problems of the people whom he ruled and willing to see in him only the puppet of forces which he could neither fully understand nor hope to control. By comparison with his two great contemporaries, moreover, his reputation has suffered no really violent fluctuations. Frederick has been seen both as a noble, patriotic, and unselfish hero and as a self-seeking, oppressive, and narrow-minded despot. Catherine, in the eyes of some contemporaries the fount of progress and enlightenment in Russia, has been presented by many more recent writers as the mere instrument of the landed ruling class. There is, by contrast, a good deal of stability and lack of fundamental change in the picture of Joseph presented both by those who wrote in his own lifetime and by most modern academic historians. Throughout it is in essentials one of a ruler of immense energy, endowed with an intense feeling of responsibility, whose sincere efforts at reform were doomed in many respects to failure by a combination of bad luck, conservative opposition, and above all his own fundamental lack of realism. Nevertheless within the limits of this essentially stable picture the emphasis has shifted perceptibly from time to time. These shifts reflect in part the advance of historical knowledge and in part the changing environment, political, social, and intellectual, within which historians and biographers of Joseph have worked.

Even during his lifetime it was clear that he was no ordinary ruler. This realization was reflected, as it had already been in the cases of Peter I of Russia and Frederick II of Prussia, in the appearance of a substantial anecdotal literature devoted to illustrating his character and ambitions.[1] Already his personal characteristics seemed clearly to set him apart from the other monarchs of Europe. His passion for work, more uncompromising even than that of Frederick II, was one of these. 'It would perhaps', wrote his first significant biographer,

---

[1] The best-known and most voluminous example is A. F. Geisler, *Skizzen aus den Karakter und Handlungen Josephs des Zweiten . . . Kaisers des Deutschen* (15 vols., Halle, 1783–91). Others are A. J. L. Du Coudray, *Anecdotes of the Emperor Joseph II during his residence in France* (London, 1778), which is a translation of a French original; and J. Pezzl, *Charakteristik Josephs II, eine historisch-biogra-phische Skizze* (Vienna, 1790). Later examples of the same type of writing are K. A. Schimmer, *Biographische Skizzen von des Geburt bis zum Tode Josephs II* (Vienna, 1844) and the collection of fifty anecdotes included in F. Schuselka (ed.), *Briefe Josephs des Zweiten* (3rd edn., Leipzig, 1846).

'be impossible to find in the distribution of his time a single hour which was not put to use, either in making plans or in carrying them out.' A generation later another student of his reign thought that 'no life has been more fully occupied than his', and yet another that he provided 'an example of indefatigable perseverance . . . which no sovereign had displayed since Peter the Great'.[1] Joseph's excessive haste, his unrealistic eagerness to put all his schemes instantly into action without consideration for the obstacles to be overcome, was another distinctive characteristic on which all the early commentators were in agreement. His passion for constant change and improvement, his desire 'to reap the harvest before it was ripe', aroused in them both admiration of his energy and public spirit and condemnation of his arbitrariness.[2] Above all he left on his contemporaries and near-contemporaries the clear impression of an unlucky man, one whose achievements had not matched his hopes and efforts. 'Great among princes and among men, he died unhappy', wrote Rioust, for 'the whole life of this prince, from his cradle to his last moments, was only a series of setbacks.' He had had the misfortune, complained Carracioli, to die just when the revolutions in France and the Austrian Netherlands were in full swing, which had affected the judgements passed on his reign: posterity would rate him higher and more fairly than his own age. Moreover circumstances had often forced him to go further than he really wished—to take part of Poland to protect himself against the growing power of Prussia and Russia; to enter the Turkish war in 1788, with disastrous results, because of an alliance with Catherine II which he could not break; to resist the Pope in order to curb the power of ultramontane forces in his own territories; to reform abuses which his mother had known of but had tolerated out of good-nature.[3]

It will be seen that these verdicts, though not uncritical, were in the main favourable. Criticism indeed was not lacking. Coxe, for example, concluded that Joseph's undoubted great qualities were counter-balanced by 'a restlessness of temper and a rage of innovation', by 'a spirit of despotism' and 'an habitual duplicity', so that his reign

---

[1] Marquis L. A. de Carracioli, *La Vie de Joseph II, Empereur d'Allemagne* (Paris, 1790), p. ix; M. N. Rioust, *Joseph II, Empereur d'Allemagne, peint par lui-même*, (2nd edn., Brussels, 1823), p. iii; W. Coxe, *History of the House of Austria from the Foundation of the Monarchy . . . to the Death of Leopold the Second* (3rd edn., London, 1847), iii. 485.

[2] Coxe, *House of Austria*, iii. 488–9; Rioust, *Joseph II*, p. iv.

[3] Rioust, *Joseph II*, p. 2; Carracioli, *Vie de Joseph II*, pp. xi, 335, 2–3.

'was a continued scene of agitation and disappointment'.[1] But the majority of the contemporary and near-contemporary judgements passed on the emperor were approving; and through many of them runs an unmistakable note of sympathy for his personal unhappiness and for the disappointment of noble hopes sincerely entertained. This sympathy and this favourable attitude continue to mark nearly all the writing on him produced in the German world during the nineteenth century. For this there are obvious explanations. Most of the writers concerned were members of the Habsburg bureaucracy, the most important middle-class element in an empire still economically backward. Moreover this official class was also the largest single element of the restricted reading public in the Habsburg territories. This group was overwhelmingly German by culture if not by blood and believed firmly in rule through a powerful, public-spirited, and centralized bureaucracy. Joseph, who had stood for such a regime using German as the language of administration, was therefore bound to make a powerful appeal to it.[2] Sometimes its admiration of him took the form of an unfair playing-down of the achievements of his mother. Maria Theresa had shown the autocratic spirit in a less uncompromising, or at least a less ostentatious, form than her son. Her efforts at administrative centralization, though real, had been less arbitrary than those of Joseph. Her intense Catholic piety and dislike of most aspects of the Enlightenment won her little regard among a nineteenth-century Austrian official class often deeply marked by distrust of papal and ultramontane influences. It was therefore tempting to see the emperor, much more than his mother, as the architect of progress in the Habsburg territories.

Thus A. J. Gross-Hoffinger, an outstanding representative of this attitude, argued in his *Historische Darstellung der Allein-Regierung Josephs des Zweiten* (Stuttgart–Leipzig, 1837), that under Maria Theresa the Austrian people had been for the most part more than a century behind civilized Europe. It was only under Joseph that they had taken 'the great step from the mental world of the sixteenth century to that of the eighteenth'. A byword for religious intolerance under the empress, Austria in Joseph's reign had become an example of toleration 'without parallel even in Protestant countries': this had meant a transformation of the intellectual life and even the judicial

---

[1] *History of the House of Austria*, iii. 541–3.

[2] See the comments of R. Bauer, 'Remarques sur l'histoire "de" ou "des" Josephismes', in P. Francastel (ed.), *Utopie et institutions* (Paris, 1963), p. 107.

system of the Habsburg provinces. Above all, 'What would the fate of Austria have been, had it remained in the condition in which Maria Theresa left it, in the storm of world events at the end of the century?' Under her the state revenues had been tiny, the army poorly organized, the administration inefficient, the people apathetic. How could this fragile structure, which could not hold its own against 40,000 Prussians in 1741, have resisted the enormously greater power of France in 1792–1809? 'All that Austria is and can yet become,' he continued, 'it has become through Joseph and will yet become through his deathless spirit.'[1] This book is one of the best statements of the Josephinist attitude to the emperor's achievements. It was an attitude widespread in the German world in the middle years of the nineteenth century. A few years after Gross-Hoffinger, for example, C. T. Heyne, in the most voluminous study of Joseph's reign yet to appear, repeated some of the same arguments almost word-for-word. Even more important than the political and physical benefits which Austria had derived from the emperor's activities, he argued, was the new spirit with which he had infused his subjects. For the traditional religious intolerance still so strong under Maria Theresa he had substituted toleration, the supreme illustration of the new and progressive attitudes which had allowed the Habsburg empire to surmount the storms which raged around it from the early 1790s onwards.[2] This adulation of Joseph was deeply unjust to Maria Thesesa and her ministers, ignoring as it did their great and permanent achievements—the administrative reorganization of 1749; the strengthening of the army and the finances; the efforts to improve the position of the peasant. It was also an attitude, significantly, hardly to be found in the non-German provinces over which Joseph had ruled.[3] Nevertheless this hero-worshipping view was to exert considerable influence down to our own day. It was to provide in particular apparent historical backing for the dream, so attractive to German liberals in central Europe, of an effectively unified Habsburg empire governed by a German–dominated bureaucracy.

[1] pp. 594–5, 601, 603.

[2] C. T. Heyne, *Joseph der Zweite, der grosse Mann des deutschen Volks* (3 vols., Leipzig, 1847), iii. 107–9, 115–16.

[3] See e.g. the Belgian view, expressed within a few years of the appearance of Gross-Hoffinger's book, in T. Juste, *Histoire du règne de l'empereur Joseph II et de la révolution belge de 1790* (2 vols., Brussels, 1845–6), which attacks his desire to create in his territories a centralized monarchy dominated by his own unrestricted will and argues that he showed '. . . the conflict between philosophy and despotism, brought together in the same man' (i. 71–2, ii. 166).

Connected with this view of the emperor was the rather different (and in its own way equally unfair) one of him as a German nationalist. No serious historian today believes that Joseph was a nationalist in any meaningful sense of the term. His efforts to make German the administrative language of his central European dominions were inspired by the desire to rationalize and centralize government, not by any idea of extending German influence or culture over non-German peoples. Yet the middle years of the nineteenth century, the period which saw the crude nationalist idealism of 1848, also saw a considerable growth of this unfounded belief among German historians. Germany, wrote Carl Ramshorn, one of the most important representatives of this point of view, was now in a period of rapid progress and above all of movement towards real unity of spirit and feeling. Joseph had been one of the great architects of this state of affairs. 'Who does not feel himself penetrated', he went on, 'at the mention of the name of this truly German man, with a feeling of holy pride that the German people can call such nobility its own?' Joseph had been 'the man who took it for his most holy duty to bring to fruition all that tended to the good of humanity, for his people and for the entire German fatherland, who planted the root of the greatness of the Austrian state and thus brought a wider citizenship to the rest of the atomized German fatherland'. Though he was willing to admit that Joseph had been sometimes over-hasty and misguided in the execution of his plans, Ramshorn refused absolutely to admit that the emperor's basic ideas and aspirations had been anything but admirable.[1] It was, however, possible to see the German nationalist Joseph, unlike the enlightened autocrat Joseph, as merely developing policies already laid down by his mother. It was through Maria Theresa and Frederick II jointly, wrote Ramshorn in a later book, that 'morality and spirituality returned to the life of Germany': the empress had been 'the noble German woman, the pride of Austria, the princess of princesses'.[2] This attitude was to have a considerable future in the twentieth century: it can be seen in the assertion of so great a scholar as Hans Schlitter that Maria Theresa and Joseph 'strove for the creation of a unified state inspired by the German spirit'.[3]

---

[1] *Kaiser Joseph II und seine Zeit* (Leipzig, 1845), pp. 6, 7, 451.

[2] *Maria Theresia und ihre Zeit* (Leipzig, 1861), pp. 3, 637.

[3] P. von Mitrofanov, *Joseph II: Seine politische und kulturelle Tätigkeit* (2 vols., Vienna–Leipzig, 1910), i, Geleitwort, p. v.

Nearly all mid-nineteenth century admiration of Joseph II was marked by a distinct tinge of romanticism. One aspect of this great movement of ideas was a taste for noble failure as against vulgar success, a deep feeling for the great man struggling with forces which overcome him but to which he is in some ultimate and unchallengeable way superior. To this feeling Joseph, through his unhappy personal life, his struggles with internal and external foes, his at least partial defeat by what could be seen as superstition and selfish conservatism, made a deep appeal. Although, or because, he achieved less than Frederick II or Catherine II, he won more sympathy as a person than either. 'In fact,' wrote Ramshorn, 'Joseph II's life is a great tragedy, rich in beautiful and truly godlike moments, but still ending like every other tragedy—with the fall of its hero, a sacrifice to his own feeling of love.'[1] 'Why,' asked a French author by no means uncritical of Joseph, 'in spite of grave mistakes, does this ruler inspire deep sympathy? It is because, under that imperial purple too often shaken by ambition, we feel none the less the beating of a heart which is sensitive, humane, generous.'[2] This is a note less often struck in the discussion of other eighteenth-century rulers; like the claims made for Joseph's liberalism and nationalism it is one which has continued to sound down to our own day.[3]

The overwhelmingly favourable attitude to Joseph which emerges from even a cursory reading of the German historical literature of the first half of the nineteenth century continued during its later decades. The factors, political and intellectual, which had generated it in the age of Metternich still operated, though with slowly decreasing effect, long after his fall. The intellectual life of the Habsburg empire continued to find in the German liberals its greatest single source of nourishment. Josephinist attitudes in church–state relations, temporarily checked by the concordat of 1855, tended to reassert themselves after 1866. Standards of historical scholarship remained lower than in any other great European state. On a popular level the result was adulation of Joseph in a series of works, several of them the products

[1] *Kaiser Joseph II*, p. 450.

[2] C. Paganel, *Histoire de Joseph II, empereur d'Allemagne* (Paris, 1845), p. 484.

[3] For a good example of a twentieth-century view of Joseph as a romantic revolutionary who, alone among the Habsburgs, could win real popular affection, see E. Benedikt, *Kaiser Joseph II, 1741-1790* (Vienna, 1936); though it should be added that this book also reproduces in its appendices a good deal of hitherto unprinted material relating to Joseph's reign.

of the centenary in 1880 of his accession to the Austrian throne.[1] On a more scholarly one the same uncritical approval and partisan spirit are unmistakable. The most complete and best-informed of the books produced to mark the centenary, by Johann Wendrinsky, shows on a larger scale the same attitude as its humbler contemporaries. Like them, and like the great bulk of the writing on Joseph during the nineteenth century, it reflects the outlook of a German liberal nationalist. The emperor's benevolence towards the people of Vienna, quite apart from his policies as a statesman, would in itself, wrote the author, 'be quite sufficient to entitle Joseph's memory to the lasting and thankful regard of posterity'. Moreover he had been the supreme embodiment of the best qualities of the German people. Frederick II had been much inferior to him as a man, a representative in his cold intelligence merely of the north German character. The emperor, by contrast, was a reflection (*Spiegelbild*) of Austrian and South German characteristics, but had been endowed with a power of work and sense of duty superior to those normal in the *leichtblutig* south. Unlike Frederick, he could write faultless German and had greatly fostered the spread of the German language and way of life.[2] Though Wendrinsky's was one of the better studies of Joseph to appear during the nineteenth century its lack of critical sense and tendency towards panegyric are depressing. Few things illustrate better than a comparison of the Austrian writing of this period on the emperor with the corresponding German discussion of Frederick II the inferiority of Austria in the more analytical aspects of intellectual life.

Outside the Habsburg dominions this uncritically favourable picture of the emperor and his work was indeed challenged. A Belgian historian pointed out that his reforms could be regarded in many respects as merely a continuation of those already embarked upon by his mother.[3] A Prussian one, probing deeper, argued that Joseph's work was inspired by the merely mechanical and rationalistic attitudes of the Enlightenment, whereas that of the great Prussian

---

[1] e.g. E. Hellmuth, *Kaiser Joseph II. Ein Buch fur's Volk* (Prague, 1862); S. Berger, *Kaiser Joseph II, sein Leben und Wirken* (Brünn, 1880); F. Babsch, *Kaiser Joseph II* (Vienna, 1880). The last, the most laudatory of all, presents him as 'a noble shoot of the glorious Habsburg ruling family', as 'a true Hero-emperor' and as 'the sublime product (*erhabene Produkt*) of the spirit of the Austrian people'.

[2] J. Wendrinsky, *Kaiser Joseph II* (Vienna, 1880), pp. 382–3.

[3] T. Juste, *Joseph II* (Verviers, 1879), pp. 12–13, 21–2.

reformers, Stein and Hardenberg, had been based on the philosophy of Kant and Fichte. This had endowed it with 'a moral–political strength of enormous extent', Protestant in origin, which expressed the finest spiritual and moral impulses of evangelical Germany.[1] Moreover the later decades of the nineteenth century saw the publication on a hitherto unknown scale of materials from the Austrian state archives. In particular a series of collections of his correspondence published from the 1860s onwards laid some of the foundations for a more balanced appraisal of Joseph and his reign.[2] Most important of all, the archivist Alfred Ritter von Arneth provided, in the ten volumes of his *Geschichte Maria Theresias* (Vienna, 1863–79), the first detailed account of the empress's reign based on unprinted materials. The book is not an analytical or comparative study; indeed it is hardly a study of any kind but rather a great collection of extracts from the documents among which Arneth spent his life. Nevertheless with all its weaknesses it showed clearly for the first time the importance of its subject. In scholarly quality as well as in size it far transcended the few previous efforts to write the biography of the empress.[3] It was now possible for the first time, on the basis of the evidence assembled by Arneth, to see Maria Theresa as a real and even bold reformer in her own right, and not as a traditionalist whose conservatism merely threw into more telling relief the radicalism of her brilliant and unfortunate son.[4] The effects of this soon began to

---

[1] O. Hintze, 'Der österreichische und der prüssische Beamtenstaat im 17. und 18. Jahrhundert', *Staat und Verfassung* (vol. i of his *Gesammelte Abhandlungen*) (Göttingen, 1962), p. 355. This famous article was first published in 1901.

[2] *Maria Theresia und Joseph II. Ihre Correspondenz sammt Briefen Joseph's an seinen Bruder Leopold*, ed. A. von Arneth (Vienna, 1867–8); *Joseph II und Katharina von Russland. Ihr Briefwechsel*, ed. A. von Arneth (Vienna, 1869); *Joseph II und Leopold von Toscana. Ihr Briefwechsel von 1781 bis 1790*, ed. A. von Arneth (Vienna, 1872); *Joseph II, Leopold II und Kaunitz. Ihr Briefwechsel*, ed. A. Beer (Vienna, 1873); *Correspondance secrète du Comte de Mercy-Argenteau avec l'empereur Joseph II*, ed. A. von Arneth (Paris, 1889–91); *Joseph II und Graf Ludwig Cobenzl. Ihr Briefwechsel*, eds. A. Beer and J. Fiedler (Vienna, 1901). Earlier publications of Joseph's correspondence had been extremely meagre. The *Lettres inédites de Joseph II, empereur d'Allemagne* (Paris, 1822), which is a translation of a German original, included only 48 letters, and the collection published by Schuselka in the 1840s only 54.

[3] e.g. K. A. Schimmer, *Die grosse Maria Theresia. Das Leben und Wirken dieser unvergesslichen Monarchin* (Vienna, 1845) which is merely a short popular work and concerned mainly with the period after 1765. The one really substantial predecessor of Arneth's book is C. Ramshorn, *Maria Theresia und ihre Zeit* (Leipzig, 1861).

[4] Arneth's discussion of the poll-tax of 1746 as an effort to introduce the

show themselves. Only two years after Arneth completed his work there appeared for perhaps the first time a book which, when making comparisons between Maria Theresa and Joseph, did full justice to the empress as against her hitherto much more highly praised successor.

This revaluation of Maria Theresa has continued in the twentieth century. The fundamental importance of the great administrative reforms in the Habsburg territories which began in 1749 were clearly understood by its first decades.[1] More recently Friedrich Walter in particular has studied them in great detail and placed their significance beyond any possible doubt.[2] In his latest work, a general discussion of the administrative history of Austria throughout modern history, he has given four times as much space to the achievements of the mother as to those of the son.[3] The more closely the two rulers are compared the clearer becomes the dependence of Joseph on the work of his mother and her ministers, the extent to which he was following in a path which they had already marked out for him. Early in the century Paul von Mitrofanov, in what remains to this day the most important scholarly book on Joseph, had fully understood this. In the government of his territories, he concluded, the emperor had merely extended the centralization and bureaucratization begun in the middle of the eighteenth century. The protectionist tariff system which he applied had existed long before his reign. Maria Theresa had aimed throughout her life at the sort of judicial reform which her son attempted. The liberation of the peasants from remaining feudal bonds, Joseph's greatest achievement in the eyes of modern writers, had begun in the 1770s with his mother still on the throne. In church-state relations the empress, though less extreme than her son, had been like him opposed to ultramontane influences. Joseph, in other words, had waged more aggressively than his predecessor a war against the remnants of the Middle Ages; but he had not originated this struggle as so many of his nineteenth-century

---

principle of equal liability to taxation of all ranks in society (*Geschichte Maria Theresias*, iv, Chap. III) is an illustration of this.

[1] See e.g. E. Guglia, *Maria Theresia. Ihr Leben und ihre Regierung* (2 vols., Münich–Berlin, 1917), especially vol. ii. This is probably still in general the most satisfactory single work on the empress.

[2] F. Walter, *Die Theresianische Staatsreform von 1749* (Vienna, 1958).

[3] F. Walter, *Österreichische Verfassungs-und Verwaltungsgeschichte von 1500–1955* (Vienna–Köln–Graz, 1972): pp. 89–108 are devoted to the work of Maria Theresa, pp. 108–14 to that of Joseph II.

admirers believed.[1] This view has been adopted by every serious writer on the subject during the last two generations. The result has been a growing refusal to regard 1780, the beginning of Joseph's personal reign, or 1765, when he became co-ruler with his mother, as real dividing lines in the history of the Habsburg territories. Increasingly it is clear that Joseph's policies simply were not different enough from those of his predecessor to justify such an attitude, that the period 1748–90, from the first great reforms of Maria Theresa to the death of her son, must be considered as a unit.[2]

The fact that Joseph's reign cannot be separated from what preceded it is underlined even more heavily by modern studies of his religious policies, the aspect of his reign which in the twentieth century, as in the nineteenth, has drawn more attention from historians than any other. Eduard Winter, in a series of books which embody a lifetime of study, has made this point with much learning. His most recent work, which sums up his conclusions, argues strongly for the existence in the Habsburg territories from the first years of the eighteenth century of an active group of reformers who wished to restrict the powers of the Catholic Church, to centralize and unify the administration, and in some cases even to improve the position of the peasantry. Moreover the conflicts between the Habsburgs and the Papacy in 1705–30, stimulated by the former's possession of territory in both north and south Italy and the resulting threat of the encirclement of the papal state, 'lead directly to the conflicts of Maria Theresa and Joseph II with the Curia'. All this meant that from the first decade of the century 'so-called Josephinism already openly makes its appearance. Its foundations were laid by Joseph I (1705–11) and not by Joseph II.' It was in the early years of the eighteenth century, not in its second half, that there took place in the Habsburg empire a transition from 'Confessional Absolutism' to 'Enlightened Absolutism'. To credit Joseph II with the establishment of a high degree of state control over the Catholic Church is therefore a positive

---

[1] P. von Mitrofanov, *Joseph II: Seine politische und kulturelle Tätigkeit*, ii. 849–50. For the author's comments on the extent of Joseph's debt to his mother in his administrative policies see i. 274 ff. See also the claim by Hans Schlitter, in his *Geleitwort* to the book, that Joseph, 'brought up simply and entirely in the school of Maria Theresa . . . expressed the policies of that princess' (i. x).

[2] See e.g. P. Müller, 'Der Aufgeklärte Despotismus in Österreich', *Bulletin of the International Commission of Historical Sciences*, ix. (1937), 25–6; R. A. Kann, *A Study in Austrian Intellectual History: From Late Baroque to Romanticism* (London, 1960), p. 137.

error; this control had already been created long before he was born.[1] One of Winter's main points, the extent to which Italy was the breeding-ground of the Josephinism which was later to show itself in central Europe, has also been developed in detail by another Austrian historian, Ferdinand Maass, in the first volume of his great collection of documents, *Der Josephinismus. Quellen zu seiner Geschichte in Österreich* (5 vols., Vienna, 1951-61). This shows the extent to which the central tenet of Josephinism, the superiority of state to church, was first put into practice in the Duchy of Milan. Indeed Maass goes further than Winter in his efforts to provide Joseph and his beliefs with a long pedigree, for he argues that the emperor's ideas were really an outgrowth of the divine right monarchy of the previous century. Joseph and his greatest minister, Prince Kaunitz, he agrees, certainly believed that the powers of the ruler were limited, both by revealed law, declared by the church, and by the natural law known by the use of the human intelligence. But these limits they thought of as fixed, as in the past, only by God, whom they continued to see as the source of monarchical authority.[2]

In other words a great deal of recent writing on Joseph II, as on other aspects of eighteenth-century Europe, has stressed continuity rather than sharp innovation. Here as in some other fields[3] there has been a marked tendency to push increasingly far back in time the origins of developments formerly thought of as new and owing little to the past. From one point of view this can be seen, with justification, as a sign of a more acute historical sense, a deeper understanding of

[1] E. Winter, *Barock, Absolutismus und Aufklärung in der Donaumonarchie* (Vienna, 1971), pp. 82-6, 91, 92, 161-2. This book carries further some of the arguments of his *Der Josefinismus. Die Geschichte des österreichischen Reformkatholizismus, 1740-1848* (Berlin, 1962), which in turn is a revised version of his *Der Josefinismus und seine Geschichte* (Brünn-München-Vienna, 1943). The same author's *Frühaufklärung: Der Kampf gegen den Konfessionalismus in Mittel-und Osteuropa und die deutsch-slawische Begegnung* (Berlin, 1966), pp. 117-36, adds much factual detail on the development of the Enlightenment in the Austrian provinces during the first half of the eighteenth century.

[2] See the summary of Maass's arguments in R. Bauer, 'Remarques sur l'histoire "du" ou "des" Josephismes', in P. Francastel, (ed.), *Utopie et institutions* (Paris, 1963), pp. 110, 112. The extent to which the Duchy of Milan developed in the eighteenth century a type of enlightened despotism of its own, parallel to but in some ways dissimilar from that to be found in Austria and drawing inspiration from Paris rather than Vienna, was first clearly brought out in a modern scholarly manner by F. Valsecchi in the second volume of his *L'Assolutismo illuminato in Austria e in Lombardia* (Bologna, 1931).

[3] See, for example, what is said about the historiography of the Enlightenment on pp. 93 ff.

how indissolubly the present is linked to the past, than the writers of the nineteenth century could boast. It may also, however, reflect increasing pessimism, rare in more hopeful generations, about the possibility of man's escaping from the past, an ebbing belief in the ability of great men of visionary ideas to set whole societies on a new path.

Finally, as elsewhere in the study of the eighteenth century, there has been a greatly increased interest during the last generation or more in the social and economic aspects of the reigns of Maria Theresa and Joseph II. Nineteenth-century writing, focused so much on personalities, on the intellectual life of Austria, and on church–state relations, paid remarkably little attention to the social background against which these rulers and their ministers had to work.[1] Recent decades, however, have seen the appearance of several important and stimulating books on this, though it is probably fair to say that these have been written in English rather than German. As long ago as 1910 Henrik Marczali provided English readers for the first time with a detailed picture of political and social realities in the most intractable of the Habsburg territories in his *Hungary in the Eighteenth Century* (Cambridge, 1910). Later Robert Kerner performed a similar function for another important area in his *Bohemia in the Eighteenth Century* (New York, 1932), though this concentrated on the position immediately after the end of Joseph's reign; and W. E. Wright has within the last decade discussed the agrarian position in the same province on a much wider time-scale.[2] In the most stimulating of all these books Ernst Wangermann has seen Joseph as helping to produce, by his reforming efforts, a considerable body of opinion, middle-class and even artisan, which aimed at really radical social and political change in the Habsburg empire. A demand for a body of fundamental law which would drastically restrict the ruler in the exercise of his powers and prevent him from behaving in an arbitrary way; a marked dislike of clerical influences and even of traditional religious beliefs; hostility to noble privileges: these marked out this group as a new phenomenon in the history of central Europe. To ward off possible papal interference and to weaken clerical and noble privileges the emperor was forced, Wangermann argues, to

[1] For example Ramshorn, *Maria Theresia und ihre Zeit*, devoted little more than seven pages (in a book of 638) to the agrarian problems of the Habsburg territories and to their economic life in general.

[2] W. E. Wright, *Serf, Seigneur and Sovereign: Agrarian Reform in Eighteenth-century Bohemia* (Minneapolis, 1966).

bring at least some sections of the huge non-privileged majority of his subjects into political life and to inspire them with hopes which he could not satisfy. The great reform of the tax system in 1789 in particular stimulated demands for the sweeping away of every aspect of the feudalism which still bulked so large throughout his dominions. Joseph thus laid, both by his reforms and by the forced retreat from some of them which marked the last months of his life, the foundations for the abortive revolutionary conspiracies which were finally suppressed in 1794 under Francis II.[1] This picture probably exaggerates the scope and weight of the popular pressures for radical reform during these years in an area still so backward in almost every way. Certainly Joseph, though he employed many non-nobles in important positions, had no desire to exclude from administrative posts members of the great noble families. Nevertheless, the difference of perspective and emphasis is striking if Wangermann's view of the forces at work under Joseph is compared with that painted by the adulatory biographers of the nineteenth century.

The effect of what has been written during the last generation or more has thus been considerably to demote the emperor from the position which he previously occupied. Instead of being seen as the single-handed originator of daring new departures he now appears as merely continuing and developing those already evolved by his predecessors. In place of the great monarch devotedly remodelling the life of his subjects in terms of the most advanced and liberal ideas we are now offered a well-meaning ruler to whose actions the social and economic structure of his dominions inevitably set limits. Once more history, as information accumulates and analysis is refined, has become depersonalized. Once more the emphasis has moved from individuals to mass interests and mass psychologies. Joseph, his faults and weaknesses seen with increasing clarity, has become a smaller man than the demigod depicted by nineteenth-century partisanship.[2] Historians have never disliked Joseph II as some of them have dis-

[1] E. Wangermann, *From Joseph II to the Jacobin Trials* (Oxford, 1959), chap. i, *passim*.

[2] Thus, for example, the most satisfactory modern study of him in English points out that in the early 1770s, when Maria Theresa hoped for the abolition of serfdom in Bohemia, he aimed merely at its amelioration; that he showed little interest in the great education ordinance of 1774, which set up the first true system of universal education in Europe; that much of his reformed criminal legislation was far from merciful; and that his attitude to the non-utilitarian aspects of higher education was distinctly hostile (P. P. Bernard, *Joseph II* (New York, 1968), pp. 52–3, 58–9, 98–100, 102–4).

liked and even despised Frederick II and Catherine II. But his historical standing has fallen rather than risen during the twentieth century. That of his mother, on the other hand, has greatly improved. Where the nineteenth century saw a brave and virtuous but essentially uncreative ruler we now see the monarch under whom the entire Habsburg administrative system was reformed, the tax system drastically overhauled, and the first successful efforts to improve the position of the peasant made. Maria Theresa appears not merely as the initiator of many of her son's policies but as the greatest ruler of the Habsburg dynasty. This revaluation is a victory, belated but real, for justice and common sense.

CHAPTER IV

GREAT BRITAIN: INDUSTRY AND PARLIAMENTARY GOVERNMENT

THE ORIGINS OF THE INDUSTRIAL REVOLUTION

IT IS a truism to say that the later decades of the eighteenth century saw a great increase in the economic strength and dynamism of Great Britain; and that this showed itself above all in a spectacular growth of industry, and of industry of a new kind. This great lurch forward was seen most clearly in a striking increase in the country's wealth and productivity; in 1830 the British gross national product was somewhere between six and nine times as great as it had been in 1700. Accompanying and underlying this were a growth of population more rapid and above all more sustained than ever before, a marked increase of urbanization and a spectacular development of internal communications. In industry there was an increase, very sharp by the standards of the past, in the use of sometimes quite complex machines; a tendency, though a less marked one, for units of production to grow bigger; and perhaps most fundamental of all a movement in the most rapidly growing sector, cotton textiles, towards a new and stricter labour discipline, and flow rather than batch production. There is now, and has for long been, general agreement by historians that this transformation constituted a turning-point not merely in British but in world history.

To expect the men who witnessed these changes to analyse them impartially, to assess their importance accurately or even to describe them fully, would be unrealistic. Their in some ways unprecedented character and the fact that their full flowering was for long confined to one or two geographical areas—Lancashire and the West Riding of Yorkshire—prevented for decades the achievement of any full or balanced view of their scope and implications. Yet it is remarkable that the great acceleration of the tempo of British economic life which developed in the last decades of the eighteenth century was not systematically described, far less explained, until a hundred years or more had passed. The shortcomings of even the best-informed contemporary writing on the subject can be clearly seen in George Chalmers's *An Historical View of the Domestic Economy of Great Britain and Ireland from the Earliest to the Present Times* (Edinburgh, 1812), perhaps the fullest work of the early nineteenth century on the recent economic history of the country. With all its considerable virtues this book is concerned overwhelmingly with what were, by the time it was published, long established as the traditional areas of economic controversy: trade, and to a lesser extent population and public revenue. Industry receives scarcely any attention. When its remarkable growth is fleetingly mentioned this is attributed simply to the effects of legislation in reducing or abolishing duties on imported raw materials, providing bounties on the export of manufactured goods, and forbidding combinations of workers. Even the explosive growth of the decade 1783–93, growth which later historians have seen as beginning a new era in the history of humanity, is dismissed with the anodyne remark that in these years 'our various manufactories were greatly promoted by the several laws which were made, year after year, for their encouragement.' Though Chalmers printed a table showing that this decade saw an unprecedented number of enclosure acts and of acts for the building of roads, bridges, canals, and harbour-works he shows no grasp of the fact that all this activity was connected, by complex chains of cause and effect, with the growth of industry.[1] Economic growth to him, as to the bulk of his contemporaries, was something to be explained in terms of government policies rather than in those of the workings of an at least partially autonomous and self-sustaining economic mechanism. In the same way and almost at the same moment Patrick Colquhoun, in his *Treatise on the Wealth, Power, and Resources of the British Empire, in every Quarter of the World*, the most ambitious statistical work of the early nineteenth century, attributed the astonishing progress of recent decades to political and constitutional, not to economic factors. 'It is fair to conclude', he wrote, 'that the rapid strides which this nation has made in the course of the last and present century towards wealth and power may fairly be imputed to the form of government, and the wisdom of its councils.'[2] Twenty years later the author of the first great work of industrial history in any language summed up the

---

[1] *Historical View*, pp. 152–4, 204, 206. This book is a revised edition of Chalmers's successful *The Comparative Strength of Great Britain*, first published in 1782.

[2] 2nd edn. (London, 1815), p. 49.

reasons for the extraordinary growth of industry in Britain as 'natural, political, and adventitious'. The country's position between north and south Europe had favoured the growth of her trade and economic strength. The rule of law, and the personal liberty and security of property which depended on it, had favoured every kind of economic progress. Finally wars, and especially the great cycle of conflict which began in 1792–3, had hampered the industrial development of Britain's competitors and given her a freer field than she would otherwise have had.[1] Few present-day writers would disagree with this argument so far as it goes; but once more the absence of any effort at explanation in strictly economic terms is striking. Britain in the later eighteenth and early nineteenth centuries was successfully resisting revolutionary political change. The idea of such change in economic life, or at least the application of the terminology of revolution to industry or trade, was thus to her more difficult of acceptance than to a society with a different political experience. It is significant that the term 'Industrial Revolution' was first used in France, at least as early as the later 1820s, and was undoubtedly inspired by analogy with the upheaval of 1789.[2]

'An epoch which has revolutionised and changed a great nation', complained an English writer on the social problems of the 1830s, 'has been going on, without any examination of its precursory events, or its ulterior influences.'[3] The complaint was justified. Throughout the Victorian age even the most detailed general histories of eighteenth-century Britain continued to give ludicrously meagre and inadequate attention to the vast economic changes which the reign of George III had witnessed. The eight volumes of Lord Mahon's *History of England from the Peace of Utrecht to the Peace of Versailles* (London, 1836–53) provided, among a mass of political and diplomatic detail, only the thinnest anecdotal account of the transformation of industry which was beginning to be visible in the 1770s and 1780s.[4] Mahon made no attempt whatever to explain why these changes occurred; indeed he showed no realization that there was anything to

[1] E. Baines, *History of the Cotton Manufacture in Great Britain* (London, 1835), pp. 88–9.

[2] Anna Bezanson, 'The early Use of the Term Industrial Revolution', *Quarterly Journal of Economics*, xxxvi (1921–2), 343–9. The first use of the term in academic writing appears to be in the *Histoire de l'économie politique* of J. A. Blanqui (Paris, 1837); see G. N. Clark, *The Idea of the Industrial Revolution* (Glasgow, 1953), p. 10.

[3] P. Gaskell, *The Manufacturing Population of England* (London, 1833), p. 33.

[4] 1858 edn., v. 1–8.

explain. Forty years later, in his *History of England in the Eighteenth Century* (London, 1878–90), W. E. H. Lecky devoted less than fifty pages (in a total of about 3,300) to a purely descriptive account of the industrial and agricultural revolutions.[1] Arkwright and Crompton seemed to him worthy of a bare mention only because their work helped to modify the styles of dress prevalent in the later eighteenth century; and though he referred in passing to 'the sudden and enormous agglomeration of population in manufacturing towns' as a new problem of the age, the phenomenon seemed to him unworthy of analysis or even of real description.

In one important and highly emotive area, however, the first decades of the nineteenth century saw the beginnings of a controversy which has smouldered and periodically erupted ever since. This was the dispute, perhaps insoluble in general terms, as to whether the Industrial Revolution had improved or degraded the material standards and indeed the whole way of life of the worker. From very early in the process of accelerated industrialization traditionalists and paternalists, especially those who tended to see in conventional religious values the cement of society and the ultimate basis of all government, can be found attacking the new factories. These were alleged to disrupt families, to break traditional social ties, to have disastrous effects on the sexual morals of those who worked in them and to destroy their physical health. This point of view, with its often strong overtones of aesthetic objection to the necessary externals of industrialization (factory chimneys and new working-class housing) was well expressed by a leading literary figure in Robert Southey's *Sir Thomas More: or, Colloquies on the Progress and Prospects of Society* (London, 1829).[2] A few years later, another writer, with the authority given by his position as a medical man, developed powerfully the moral arguments against 'the separation of families, the breaking-up of households, the disruption of all those ties which link man's heart to the better portion of his nature' which he held industrialization to involve. The point was driven home by painting an idyllic picture of the handloom weaver in the good years of the 1790s, when he was 'kept apart from all associations which might injure his moral worth' and comparing it with the physical and

---

[1] New edn. (London, 1892), vii. 173, 189, 241–90.

[2] See e.g. the attacks on industrialism and the unbridled competition which Southey associated with it in the second edition of the book (London, 1831), i. 170, 197, ii. 157–8, 161–3.

spiritual degradation which had overtaken him since he had been reduced to a factory hand.[1]

On the other side of the argument Macaulay, in a review of Southey's book, satirized objections to modern industry, in a memorable phrase, as a preference for 'rose-bushes and poor-rates, rather than steam-engines and independence'. He went on to claim that industrialization had greatly improved general living standards and to complain that the attractiveness of the life of the average man before its advent had been grossly exaggerated. 'Does Mr. Southey think', he asked, with much point, 'that the body of the English peasantry live, or ever lived, in substantial or ornamented cottages, with box-hedges, flower-gardens, beehives, and orchards?'[2] The battle between the two sides had been fairly joined; and though it goes on to this day it cannot be said that either has achieved or is likely to achieve decisive victory.[3]

Well before the middle of the nineteenth century, however, the tendency to discuss and evaluate the Industrial Revolution in social rather than economic terms, in those of social policy rather than economic forces and the more or less accurate measurement of economic aggregates, had taken root in Britain. During the second

[1] P. Gaskell, *The Manufacturing Population of England*, pp. 7–8; *Prospects of Industry: being a Brief Exposition of the Past and Present Conditions of the Labouring Classes* (London, 1835), p. 18.

[2] *Critical and Historical Essays* (London, 1883), pp. 106, 118. The reference is to *Sir Thomas More*, 2nd edn., i. 173–4. The most systematic and best-informed defence of the factory system in the first half of the nineteenth century is that in Baines, *History of the Cotton Manufacture*, chap. xvi.

[3] Since the 'standard-of-living controversy' is not relevant to the question of the origins of the Industrial Revolution and is concerned essentially with the period 1790–1850, which is outside the chronological scope of this book, it will not be further discussed here. Condemnation of the social and moral evils of industrialism and a rosy view of the life led by the ordinary worker, at least until the middle of the eighteenth century, continued to be the dominant note of most comment in the later nineteenth and early twentieth centuries (e.g. A. Toynbee, *Lectures on the Industrial Revolution in England* (London, 1884), pp. 58, 69, 71; C. Beard, *The Industrial Revolution* (London, 1901), pp. 2–3, 8–9, 16). It was still sounding loudly in the 1940s in such works as C. Clark, *Conditions of Economic Progress* (London, 1940) and J. Kuczynski, *A Short History of Labour Conditions under Industrial Capitalism* (London, 1942–45), vol. i. Professor E. J. Hobsbawm has continued to make it heard in two important articles, 'The British Standard of Living, 1790–1850', *Economic History Review*, 2nd ser., x, No. 1 (1957); and 'The Standard of Living during the Industrial Revolution', *ibid.* xvi, No. 1 (1963). The other side of the argument has been ably put by T. S. Ashton, 'The Standard of Life of Workers in England, 1790–1830', in F. A. Hayek (ed.), *Capitalism and the Historians* (Chicago, 1954); and by R. M. Hartwell, 'The Rising Standard of Living in England, 1800–1850', *Economic History Review*, 2nd ser., xiii, No. 3 (1961).

half of the century, indeed until the First World War, it continued to be the dominant one in writing on the subject. This is nowhere more evident than in the series of lectures in which, in 1881–2, Arnold Toynbee provided the first attempt at a systematic general discussion of the subject.[1] Neither of the two most important intellectual influences working on Toynbee, the aesthetics of John Ruskin and the idealist political philosophy of T. H. Green, were at all likely to encourage an analytical economist's view of the Industrial Revolution; and in fact the lectures contain, by present-day standards, extraordinarily little on how economic change in fact happened. Toynbee had nothing to say on the financing of the new machinery and methods, or on the details of productive processes or the fluctuations of prices, wages, population, or interest-rates. Economic analysis in any modern sense of the term simply did not enter into the picture he drew. He was concerned above all with ideas on social policy as embodied in government action and legislation. He saw the Industrial Revolution essentially in terms of the defeat of a system of state regulation of economic life, in the main well-meaning and far from ineffective, by the dynamic but in many ways destructive new ideal of free competition. 'The essence of the Industrial Revolution', he claimed, 'is the substitution of competition for the mediaeval regulations which had previously controlled the production and distribution of wealth.'[2] From this standpoint it was not difficult for him to conclude that the agricultural labourer had been deeply injured by enclosures and the industrial worker by fluctuations in the demand for labour, and therefore in employment and wage-rates, of a severity previously unknown. Poverty and pauperism had thus grown side by side with a vast increase in production and wealth. Toynbee has been accused with justice of 'tracing back the origin of social evils to fallacious economics and hoping for a remedy from a change of mind'.[3] With all its faults, however, his work had a deep and lasting influence. More than any other, it formed opinion on the subject during the generation after its publication; and its influence lasted, even among professional academics, at least down to the 1920s.[4] The most important large-scale work of economic

---

[1] *Lectures on the Industrial Revolution in England* (London, 1884).

[2] *Lectures*, p. 85.

[3] Clark, *The Idea of the Industrial Revolution*, p. 31.

[4] On the reasons for the influence of Toynbee's work see H. L. Beales, 'The Industrial Revolution', *History*, xiv (1929–30), 125. The book was reprinted in the United States as recently as 1956.

history produced in Britain during the nineteenth century, Archdeacon William Cunningham's *The Growth of English Industry and Commerce in Modern Times* (Cambridge, 1882) shows, in a rather less extreme and less moralistic form, the same attitudes. It reveals the same sympathy with the interventionist and regulatory tendencies of the seventeenth and early eighteenth centuries, the same dislike of 'reckless competition' and its effects on the ordinary worker, and the same condemnation of economic fluctuations and instability, which are seen as an inevitable result of uncontrolled industrial growth.[1]

In 1906 the intellectual level of the entire discussion was raised by the publication of what remains to this day the most detailed study of the subject, *La Révolution industrielle au dix-huitième siècle* by the French historian Paul Mantoux. This remarkable book did not merely assemble a greater wealth of material and of factual detail than Toynbee or Cunningham had achieved. More important, it also attempted, for the first time on a significant scale, to explain the great transformation of much of the British economy during the later eighteenth century in essentially economic terms. To Mantoux expanding trade and the growing market for manufactured goods which it generated made the growth of industry and the adoption of new machinery inevitable. Simultaneously improvements in agriculture made it possible to feed the increasing population which provided the labour-force needed to man the new factories. The expansion of trade and credit, with its reactions on the organization and technology of industry, and the growth of food production, were thus the twin foundations upon which a structure of dramatic industrial growth had been raised.[2] This was a much more workmanlike and more modern analysis than any hitherto achieved by a British historian; but it is significant that in spite of its great virtues the book was not translated into English until 1928.

It was not until after the First World War that there existed in Britain an intellectual climate which favoured the production of studies of the subject giving adequate weight to the economic and material factors involved as well as to the political and intellectual ones. For this delayed development there were a number of reasons.

[1] *The Industrial Revolution* (Cambridge, 1908), pp. 619–20, 628 ff., 667, 668 ff. This book is a reprinting of the later sections of *The Growth of English Industry and Commerce*.

[2] *The Industrial Revolution in the Eighteenth Century* (London, 1928), pp. 137, 190, 487.

Probably the most important was that the neo-classical economists dominant in the later nineteenth and early twentieth centuries—Jevons, Marshall, Menger, Walras, and others—were interested primarily in the functioning of the market mechanism in the short term. They paid little attention to the long-term dynamics of economic systems. Indeed the widespread belief in the decades before 1914 in the inevitability of progress meant that economic growth scarcely appeared as a real problem. Attention was therefore concentrated on increasingly sophisticated 'micro-economic' analysis rather than on the study of 'macro-economic' aggregates over long periods. The 'Historical School' of economists, led by such figures as Roscher and Schmoller, which flourished in Germany during the later decades of the nineteenth century, was an exception to this generalization. So in a different way were the Marxists. But neither of these groups had much influence on the highly insular intellectual atmosphere of Great Britain.[1] Since the 1920s, however, a series of far-reaching changes has transformed the position. The First World War undermined a whole system of assumptions still tenable in 1914. The emergence of the underdeveloped countries as a political force since the Second World War, coupled with an increasingly direct challenge to capitalism from various forms of collectivism, has greatly stimulated interest in the whole subject of economic growth. Economic theory, notably in the hands of Keynes and his followers, has adapted to the new situation. The result has been the rewriting of the history of the Industrial Revolution in Britain with a wealth of detailed analysis hitherto unknown and with intellectual standards of a rigour inconceivable to Toynbee or Cunningham.[2]

Here as elsewhere, rising intellectual standards in the writing of history have meant increasing complexity, the introduction into a picture previously simple and sharp-edged of a multitude of new lights and shades. Expanding knowledge has stimulated the growth of doubt and the erosion of old certainties, while offering scope for new disagreements. Nowhere has this been more apparent than in the

[1] Schmoller's *The Mercantile System and its Historical Influence illustrated chiefly from Prussian History* was published in an English translation in 1884; but it was not really concerned with industrial growth of the kind seen in England during the Industrial Revolution.

[2] This paragraph draws heavily on the editor's introduction to *Science, Technology and Economic Growth in the Eighteenth Century* ed. A. E. Musson (London, 1972). See also R. M. Hartwell, 'The Causes of the Industrial Revolution: An Essay in Methodology', *Economic History Review*, 2nd ser., xviii, No. 1 (1965), 165–6.

chronology of the subject. Until the third decade of this century it was generally agreed that the beginnings of the Industrial Revolution could be dated more or less precisely to the first years of the reign of George III. Moreover it was common ground that this great process of economic change had begun suddenly, with little warning or visible preparation. 'Previously to 1760,' wrote Toynbee, 'the old industrial system obtained in England; none of the great mechanical inventions had been introduced; the agrarian changes were still in the future.' Cunningham, with somewhat greater caution, believed that 'Despite the gradual economic development, it seems likely enough that, while centuries passed, there was little alteration in the general aspect of England; but the whole face of the country was changed by the Industrial Revolution.' Other writers were less hesitant. 'England of the first part of the eighteenth century', claimed one, 'was virtually a mediaeval England, quiet, primeval, and undisturbed by the roar of trade and commerce. Suddenly, almost like a thunderbolt from a clear sky, were ushered in the storm and stress of the Industrial Revolution.' Another had no doubt that 'The change . . . was sudden and violent. The great inventions were all made in a comparatively short space of time. . . . In a little more than twenty years all the great inventions of Watt, Arkwright and Boulton had been completed and the modern factory system had begun.'[1]

Judgements of this kind were inevitable in an age when the economic history of England during the sixteenth and seventeenth centuries, or even during the first half of the eighteenth century, had still received little serious attention. They reflect a tendency inherent in most ordinary human beings, that to associate change with more or less precise dates or short periods of time, with specific events and with identifiable individuals. They parallel in the economic sphere the readiness of many nineteenth-century writers to identify the Enlightenment in France with Voltaire or the *Encyclopédie* and to see Joseph II as the sole creator of hostility to clericalism in the Habsburg Empire.[2] Fifty years of increasingly detailed and expert research has now made attitudes of this kind untenable so far as the Industrial Revolution is concerned. The idea of it as a clearly

---

[1] Toynbee, *Lectures*, p. 32; Cunningham, *The Industrial Revolution*, p. 613; Beard, *The Industrial Revolution*, p. 23; H. de B. Gibbins, *Industry in England: Historical Outlines* (10th edn., London, 1920), p. 341. The last mentioned was first published in 1896.

[2] See above, pp. 82, 179.

identifiable historical entity, one with a well-defined beginning and end (conventionally placed somewhere about 1760 and 1830) has become more and more difficult, perhaps impossible, to sustain. In one direction it has been projected forward, as a continuing process lasting to the present day. In another it has been pushed further and further back in time in an effort to explain its origins: the great changes of the later eighteenth century have been seen as foreshadowed and prepared for by developments generations, even centuries, earlier. To the pre-Enlightenment and pre-Romanticism with which historians have enriched and confused our view of the past[1] must therefore be added the pre-Industrial Revolution. The result is that, in Sir George Clark's words, 'From the chronological point of view the idea of the Industrial Revolution has collapsed.'[2] Even if this judgement may show too great a readiness to dismiss completely the traditional presuppositions, they have been gravely weakened and are now increasingly impossible to justify.[3]

Was there in fact, as Toynbee and his followers assumed, only one 'industrial revolution', that of later eighteenth-century Britain, deserving of the title? In a controversial article of 1934 Professor J. U. Nef argued that the England of Elizabeth and the early Stuarts had seen technological innovation and industrial growth which justified the use of the term, and that 'The rise of industrialism in Great Britain, therefore, can be more properly regarded as a long process stretching back to the middle of the sixteenth century and coming down to the final triumph of the industrial state towards the end of the nineteenth, than as a sudden process associated with the late eighteenth and early nineteenth centuries.'[4] From a less extreme but not totally dissimilar point of view Professor F. J. Fisher has seen in the England of the seventeenth century economic growth and social change which may have been an essential precondition for the transformation which began in the later eighteenth century.[5]

[1] See above pp. 93 ff.

[2] *The Idea of the Industrial Revolution*, p. 29.

[3] E. E. Lampard, *Industrial Revolution: Interpretations and Perspectives* (Washington, Service Centre for Teachers of History, 1957), pp. 4–5, provides a good selection of differing opinions from historians as to whether the idea of a distinct Industrial Revolution in the later eighteenth century is still tenable or not.

[4] 'The Progress of Technology and the Growth of Large-scale Industry in Great Britain', *Economic History Review*, v (1934), 3–24. See also Nef's *The Rise of the British Coal Industry* (London, 1932), i. 165.

[5] F. J. Fisher, 'The Sixteenth and Seventeenth Centuries: The Dark Ages in English Economic History', *Economica*, New Series, xxiv (1957), 2–18.

Moreover the allegedly unique character of the late eighteenth-century transformation has, in the last generation or more, come under increasingly heavy fire. J. A. Schumpeter, perhaps the most ambitious analyst of economic phenomena whom the world has hitherto produced, explicitly rejected the idea of the Industrial Revolution of the later eighteenth century as a 'unique event or series of events that created a new economic order' and argued that it was 'on a par with at least two similar events which preceded it and at least two more which followed it'.[1] It has even been claimed that the introduction of the fulling-mill into the English wool industry in the thirteenth century constituted a kind of medieval industrial revolution.[2] The traditional concept of unique and suddenly-transforming changes which took place abruptly and with little warning has also been undermined by a realization that in most industries progress was the result, not of rapid and revolutionary change but of a series of innovations, spread over a relatively long period, which maintained or increased the rate of growth.[3] As the attention of historians has ceased to be focused so exclusively as in the past on the industries in which truly revolutionary transformation can be seen (cotton textiles and to a lesser extent iron) and as less spectacular but still important ones (paper, glass, the processing industries in general) have gained a greater share of the limelight, this reversal of old attitudes has become increasingly convincing.

There is obviously a point beyond which this merging of the industrial Revolution of the later eighteenth century with its antecedents, this smoothing of rough edges and denial of uniqueness, cannot be pushed. No one doubts that at the start of the 1780s many aspects of the British economy began to grow at an unprecedented rate and that this truly marked the beginning of a new era. Nef sees the 1780s as the decade in which Britain, at least in economic terms, broke away from 'a homogeneous and relatively stable eighteenth-century civilization'. It seems certain that from about 1780 British industry began for the first time in history to experience a sustained growth of output at an annual rate of two per cent or more; and one of the most challenging recent books on the dynamics of economic develop-

[1] *Business Cycles* (New York, 1939), i. 253.
[2] E. M. Carus-Wilson, 'An Industrial Revolution of the Thirteenth Century', *Economic History Review*, ix (1941), 39–60.
[3] For a good example of this argument, based largely on the case of the paper industry, see D. C. Coleman, 'Industrial Growth and Industrial Revolution', *Economica*, New Series, xxiii (1956), 1–22.

ment has defined the two decades 1783–1802 as 'the great watershed in the life of modern societies', since these years saw in Britain, for the first time anywhere in the world, a 'take-off' into sustained growth of a truly modern kind.[1] The Industrial Revolution therefore still retains some revolutionary characteristics of which it can never be deprived. But the clarities and simplicities of Toynbee and Cunningham have gone for ever. Research and the accumulation of knowledge have generated constructive doubt and undermined certainties.

Other traditional assumptions than the chronological ones have come under attack. What, for example, was the connection of the Industrial Revolution with capitalism? Did it, as Toynbee and Cunningham assumed, involve a great extension and intensification of capitalist relationships, or was it in this respect relatively unimportant by comparison with the growth of these relationships in earlier centuries?[2] Again, how far can industrialization be necessarily linked with political and social modernization? Are all the obvious aspects of the latter—the choice of holders of administrative posts by some form of competitive process; the assignment to public officials of relatively specific functions; an increase in geographical and social mobility; a high degree of literacy; a decrease in the importance of family links; a growth in the importance of towns as against the countryside—possible without modern industry? Was the eighteenth-century Dutch Republic, for example, a society which had largely modernized itself without being industrialized?[3] These are far-ranging questions which cannot be treated here. But the fact that they, and others like them, are attracting more and more attention shows how views of the Industrial Revolution and its significance have widened in scope and increased in sophistication in the last generation.

Given that there was a spectacular quickening of the pace of industrial growth in Britain in the later eighteenth century (whether or not this quickening justifies the use of the unsatisfactory term

---

[1] J. U. Nef, 'The Industrial Revolution Reconsidered', *Journal of Economic History*, iii (1943), 6; W. G. Hoffmann, *British Industry, 1700–1950* (Oxford, 1955), p. 30; W. W. Rostow, *Stages of Economic Growth* (Cambridge, 1964), p.38.

[2] See C. H. George, 'The Making of the English Bourgeoisie, 1500–1750', *Science and Society*, xxxv, No. 4 (Winter 1971), for a Marxist development of the argument that 'Our world is far more qualitatively defined by capitalist relations than it is by machines' (p. 391), and that therefore the industrial changes of the eighteenth century were much less important in the creation of the modern world than the growth of capitalism in the sixteenth and seventeenth centuries.

[3] See E. A. Wrigley, 'Le Processus de modernisation et la révolution industrielle en Angleterre', *Annales*, 28ᵉ année, No. 2 (March–April 1973), 519–40.

'revolution') why did this happen? Clearly some weight must be given, in attempting to answer this question, to factors which are not strictly economic. A number of writers have seen as important the development in Britain of an intellectual environment favourable to change, and above all the diffusion of scientific ideas and more scientific attitudes to economic organization. This, it is argued, aided powerfully the invention and adoption of new machinery and processes. Undoubtedly, although the Industrial Revolution was very much more than a mere discovery of a series of new machines, mechanical invention was a major and probably crucial component in it. Contemporaries were well aware that innovations such as the silk-throwing mill set up by Thomas Lombe at Derby in 1719, one of the wonders of the age, or later the new canals, were technological 'breakthroughs' of an unprecedented kind.[1] 'What distinguished Britain from the rest as the eighteenth century wore on', writes a recent analyst, 'was the scale of the inventive effort that went into the breaking of crucial technical bottlenecks, and the scale of the entrepreneurial corps which introduced them [*sic*] as the century moved towards its close.'[2] In this view, the scientific revolution laid part of the foundations for the industrial one by stimulating a permanent input of new inventions and technology into the economy. There has been a marked tendency in recent writing to stress the connections between science and technology in Britain during this period and to abandon any idea of new techniques as in the main the creation of uneducated and even illiterate artisans.[3] Even where technological advance was the work of men with little formal education it was very often backed by systematic thinking of an essentially scientific kind, while scientific knowledge and scientific attitudes had penetrated deeply into industrial society.[4] Yet to attribute the great economic leap forward which Britain experienced in the later eighteenth century mainly to intellectual influences of this kind is quite unconvincing. France, after all, produced more scientific discoveries than Britain in this

[1] The point is well made in the numerous comments quoted in Witt Bowden, *Industrial Society in England towards the End of the Eighteenth Century* (New York, 1925).

[2] W. W. Rostow, 'The Beginnings of Modern Growth in Europe: An Essay in Synthesis', *Journal of Economic History*, xxxiii (1973), 571.

[3] See, for example, T. S. Ashton, *The Industrial Revolution, 1760–1830* (London, 1958), pp. 16–21; and especially A. E. Musson and E. Robinson, *Science and Technology in the Industrial Revolution*, chap. iii.

[4] A. E. Musson and E. Robinson, 'Science and Industry in the Late Eighteenth Century', *Economic History Review*, 2nd ser., xiii (1960), 222–44.

period. Far more scientific work related to industry was done in France than on the other side of the Channel; and in France this type of work received far more official encouragement. Yet by the 1780s, and probably earlier, Britain had achieved in industrial technology a clear lead over her neighbour. The lack here of any correlation between degrees of scientific and technological achievement is striking.[1]

It can also be argued that Britain innovated and advanced because she possessed by the eighteenth century a more open and flexible social structure than any other great European state, one in which vertical social mobility was easier than elsewhere and the social prestige of business higher.[2] Yet contrasts of this kind can be easily exaggerated. In France, which again forms the most obvious basis for comparison, neither guild restrictions as an obstacle to economic and technological change nor the indifference of the ruling class to business opportunities were as great as is often supposed by English-speaking historians.[3] As yet we simply do not know enough about the relationship between social structure and economic change in eighteenth-century Britain to do more than rather loosely assume or postulate some general causal connection.[4]

If, therefore, the transformation which began to affect at least parts of British industry in the later eighteenth century cannot be satisfactorily explained in terms of intellectual and sociological factors any adequate explanations must be couched largely in economic ones. From this conclusion few historians would now dissent. 'Economic', however, is something of an umbrella term under which many differing forces can be grouped; and there has been a marked tendency in most recent writing to concentrate on a single variable, to exaggerate its effects, and attribute to it a result which was achieved by a number of such factors combined.

For a century the growth of her overseas trade has been perhaps the most favourite single explanation for the accelerated industrial

[1] See P. Mathias, 'Who unbound Prometheus?', reprinted in A. E. Musson (ed.), *Science, Technology, and Economic Growth*, pp. 79–82; and F. Crouzet, 'England and France in the Eighteenth Century: A Comparative Analysis of Two Economic Growths', reprinted in R. M. Hartwell (ed.), *The Causes of the Industrial Revolution in England* (London, 1967), pp. 160–1.

[2] T. S. Ashton, *An Economic History of England: The 18th Century* (London, 1955), pp. 20–1; D. S. Landes, 'Encore le problème de la révolution industrielle en Angleterre', *Bulletin de la société d'histoire moderne*, 12th ser., vol. 18, p. 7.

[3] Crouzet, 'England and France', pp. 157–60.

[4] Hartwell, 'The Causes of the Industrial Revolution', 176–7.

growth which Britain saw in the later eighteenth century. This trade, and particularly that with the colonies, it is argued, generated change in two main ways. The profits gained from it were the main source of plentiful supplies of capital for investment in new machines and processes. At the same time growing overseas markets powerfully increased the demand for British manufactured goods. This double stimulus, it is claimed, was of decisive importance. The primacy of overseas trade as a cause of economic change in the later eighteenth century was accepted almost without question by both Toynbee and Cunningham. Around 1760, wrote the former regretfully, 'The old simple conditions of production and exchange were on the eve of disappearance before the all-corroding force of foreign trade', while the latter concluded flatly that 'The Industrial Revolution had been occasioned by the commercial expansion of the earlier part of the eighteenth century.'[1] The influence of this trade in increasing the demand for Birtish goods and in making that demand more broadly based and therefore less subject to dislocations and fluctuations than a merely domestic one could have been, has continued to appear fundamental to many historians down to the present day. This expanded and relatively secure market which the old methods were incapable of satisfying, it is argued, powerfully encouraged the adoption of new productive techniques. Moreover, overseas commerce stimulated directly the growth of shipbuilding and its ancillary trades and also that of the processing industries which depended on imported raw materials such as sugar and tobacco. Indirectly it helped to increase the proportion of the population which lived in cities. Access to overseas markets and the low cost of transport by sea gave the export industries the advantage of important economies of scale, so that they and those which supplied them with raw materials had higher growth rates than the economy in general.[2] Recently an influential author has claimed that in 1700–50 the output of British industries producing merely for the home market increased by only 7 per cent while that of those producing for export grew by 76 per cent: for the decades 1750–80 the corresponding percentages were 7 and 80.[3]

[1] Toynbee, *Lectures*, p. 56; Cunningham, *The Industrial Revolution*, p. 668.
[2] See, e.g. Bowden, *Industrial Society*, pp. 65–8; F. C. Dietz, *The Industrial Revolution* (London, 1930), pp. 23–4; Crouzet, 'England and France', p. 166; Rostow, 'The Beginnings of Modern Growth', 556–60.
[3] E. J. Hobsbawm, *Industry and Empire* (Pelican edn., 1969), p. 48.

All this seems to constitute a formidable case. Yet it is one by no means immune to challenge. In the first place Britain's exports, however rapidly they may have been growing, amounted in the eighteenth century only to about 5–7 per cent of her gross national product. It was the growth of domestic rather than of foreign demand which was really significant for British economic growth, at least down to about 1780. A rising population, a growth of real wages, and a series of good harvests after 1730, all combined to produce a large and permanent widening of the home market for manufactured goods. Moreover a flourishing foreign trade, even if it were a necessary condition of an industrial revolution (which is far from clear), was by no means a sufficient one. The Dutch Republic in 1700 had in proportion to its population imports and exports five or six times as great as those of Britain and yet a later industrial experience totally different from that of its larger neighbour. Certainly the mere possession of colonies, even enormous and potentially very valuable ones, did not necessarily add anything to a state's industrial potentialities in the eighteenth century, as the experience of Spain shows very clearly. Again, did external trade, as has so often been asserted, contribute decisively to capital formation in Britain before and during the Industrial Revolution? A study of the subject which is perhaps the most penetrating so far produced suggests that in the period 1700–80 about 6–8 per cent of the total was in fact generated by external trade: even if the analysis is confined to the decades 1760–90 the proportion is only 7–9 per cent. Neither the navy nor the merchant fleet had more than marginal significance as consumers of British products (metals, sailcloth, etc.) and hence as a direct stimulus to the growth of industry. In sum, the probable maximum effect of Britain's foreign trade in the eighteenth century was to increase her total productive strength merely by something between five and eleven per cent.[1] No doubt these estimates can and will be challenged. But the assumption, with the apologetic and self-condemnatory overtones which it can easily assume, that Britain was able to finance her industrial growth in some sense at the expense of foreigners, or by exploiting her colonies, is now more difficult than ever before to sustain.

Was capital, and in particular 'fixed' capital in the form of factories, machines, etc., in any case such a critical factor of production in the

[1] P. Bairoch, 'Commerce international et genèse de la révolution industrielle anglaise', *Annales*, 28ᵉ année, No. 2 (March–April 1973), 541–71.

eighteenth century as the emphasis laid upon it by many historians would tend to suggest? The tradition of attaching great importance to it stems from Adam Smith and the classical economists; and the prominence they gave to the 'primary' accumulation of capital as a means of providing the resources needed to finance industrial growth was later echoed by Marxist theory. But there seems little reason to think that eighteenth-century Britain was ever short of capital for use in industry. The amount of investment during the period in factories, machinery, even canals, was by modern standards (and indeed even by those of that age) remarkably small. The wars of 1793–1815 imposed a terrifying £1,000m. in direct military and naval costs on the country. By comparison the canal system developed since 1750 had by the time of Waterloo needed only about £20m. to build.[1] The amount of 'fixed' capital needed by industry in an age when technology was still simple and productive units small was surprisingly limited.

Foreign trade and the accumulation of capital have always been the most popular single explanations of Britain's ability to transform parts of her industrial structure in the later eighteenth century. But many other factors have been put forward as contributing to this achievement. Population growth is one of these. It has been suggested that the progress of technical innovation from the 1760s onwards 'was itself stimulated by the great upsurge of population which began a generation before'.[2] But though a demographic change of this kind may well have been a necessary condition of rapid industrial change it too was clearly not a sufficient one. An 'upsurge of population' is visible over much of western Europe from the middle of the eighteenth century onwards: only in Britain was it followed by an industrial revolution. In any case the areas of northern England which were to be the birthplace of the new methods and organization suffered from having too few rather than too many inhabitants, from a shortage rather than an oversupply of labour. The need to reduce

---

[1] This pragraph depends heavily upon the study by P. Mathias, 'Capital, Credit and Enterprise in the Industrial Revolution', *Journal of European Economic History*, ii. No. 1 (1973), 121–43, especially pp. 122–4, 126–8. Mathias's view contradicts that of S. Pollard, 'Investment, Consumption and the Industrial Revolution', *Economic History Review*, 2nd ser., ix. No. 2 (Dec. 1958), 215–6, and 'Fixed Capital in the Industrial Revolution', *Journal of Economic History*, xxxiv (1964).

[2] Phyllis Deane and W. A. Cole, *British Economic Growth, 1688–1959: Trends and Structure* (Cambridge, 1967), p. 97.

if possible high labour costs may well have been the most powerful of all incentives to invent and use new labour-saving machines.[1]

More specific explanations of Britain's unique experience have also been offered. It has been powerfully argued by Professor Ashton that the tendency for interest-rates to fall and for capital thus to become available more cheaply was the most important single reason why the pace of economic advance quickened after the middle of the century. The forces making for economic change and growth, he argues, were notably stronger in periods when interest-rates were low. Cheap money tended to encourage the enclosure of common lands, the building of canals and turnpikes, and the construction and use of new machines; peaks of activity in these respects broadly coincided with the years when the cost of borrowing money was low. Even the process of invention was responsive to the same stimulus. Years when large numbers of new patents were registered were almost always also years of cheap money.[2] It is uncertain, however, how much importance should be attached to the cost of credit as an element in the surprisingly complex and in many ways still obscure economic picture of eighteenth-century Britain. The long-term trend of interest-rates during the period was certainly downwards; and naturally business men were often influenced in their plans and decisions by the price to be paid for borrowed money. But the existence of marked regional differences in interest-rates, as well as the obvious distinction between those for short and long-term loans, complicates the situation. The most detailed and expert study of the question is driven to conclude that 'The significance of interest-rates for industry and trade is one of the great unknowns of eighteenth-century history', and that 'There is no case for asserting the primacy of interest over other costs' in the making of business decisions during the period.[3]

Another, though less plausible, explanation of the great innovations of the late reighteenth century in terms of a specific economic factor is the argument that the pace of change was accelerated by profit inflation. Wages, it is claimed, lagged considerably behind the rising

---

[1] Crouzet, 'England and France', pp. 170–1.

[2] *The Industrial Revolution*, p. 11; *An Economic History of England: The 18th Century*, pp. 41, 45, 89, 108.

[3] L. S. Pressnell, 'The Rate of Interest in the Eighteenth Century', *Studies in the Industrial Revolution Presented to T. S. Ashton* (London, 1960), especially pp. 179–81, 195 ff., 210.

prices of manufactured goods. The growth of population, Irish immigration, and the effects of enclosures in displacing labour from the land to the manufacturing centres, helped to keep the price of labour low. This meant large windfall profits for the capitalist, particularly as the rents he had to pay for land and mining rights also tended to lag behind the prices he could charge. Large profits meant large investment in new machines; and this method of financing new processes was particularly important since bankers were often unwilling to sink money in enterprises of this kind. Without profit inflation, the main advocate of this view admits, there would still have been great changes; but 'the industrial progress would hardly have been revolutionary in character'.[1] This general line of argument had already been given the blessing of no less an economist than J. M. Keynes, who argued in 1930, in his *Treatise on Money*, that 'the wealth of nations is enriched not during Income inflations, but during Profit inflations—at times, that is to say, when prices are running away from costs'. But the idea of profit inflation as a significant element in the Industrial Revolution has not won the adherence of economic historians. It seems that in fact money wages rose more, rather than less, than the prices of manufactured goods in the second half of the eighteenth century; and the whole idea of a wage-lag has been subjected to searching attack in terms of economic theory.[2] Nor does this by any means exhaust the list of isolated economic variables whose individual contribution to the coming of modern industry has been stressed by historians. Before the First World War a pioneering writer, W. J. Ashley, argued that the increasing productivity of agriculture had been both the prerequisite and, through the increase in population which it made possible, the promoter, of industrialization. More recently a case for the key position of agriculture in the preparation for industrial 'take-off' has been made once more by Professor A. H. John.[3]

One thing seems clear. No explanation of the economic changes

[1] E. J. Hamilton, 'Profit Inflation and the Industrial Revolution, 1751–1800', *Quarterly Journal of Economics*, lvi (1942), 256–73.

[2] See the brief discussion of the profit inflation theory in R. M. Hartwell, 'The Causes of the Industrial Revolution: An Essay in Methodology', 173–4.

[3] *The Economic Organization of England* (London, 1914), p. 136; 'Aspects of English Economic Growth in the First Half of the Eighteenth Century', *Essays in Economic History*, ed. E. M. Carus-Wilson, ii (London, 1962), especially pp. 364–6, 371.

in later eighteenth century Britain entirely or mainly in terms of a single factor—foreign trade, capital formation, population growth, rates of interest or profit—is satisfactory. The picture at present available is in the first place that of a slow process of relatively balanced growth with roots deep in the past, which lasted until the 1760s or even the 1780s. This growth, the product of a variety of economic and non-economic factors—changing social attitudes and the extension of education; widening foreign markets; accumulating supplies of capital and growing population; technical innovation and improving transport—then accelerated in the later decades of the century and became increasingly 'unbalanced' as some sectors of the economy (iron and cotton textiles) developed explosively. This very rapid growth stimulated other branches of industry and in particular increased the pressures making for technical change in the economy at large.[1] This picture is essentially the product of work done in the last thirty years or less. Whereas current views of some of the other problems discussed in this book are still deeply influenced by writing of the nineteenth century, some of which has remained fresh and significant down to the present day, economic history, strikingly neglected until our own times, owes little such debt. No nineteenth-century writer influences our view of the Industrial Revolution in Britain as Tocqueville does our thinking about the old regime in France or perhaps Onno Klopp or Treitschke our attitude to Frederick II. By comparison with other branches of historical study economic history (and not merely that of the eighteenth century) has the advantages of youth and freedom from too intimidating a legacy of past scholarship. The study of the Industrial Revolution is now benefiting considerably from work on the problems of underdeveloped countries in the present day, and from the insights which this has given. In the future it will almost certainly gain even more from the work of anthropologists, sociologists, and social psychologists, so that increasingly the idea of 'patterns of development' will carry the discussion beyond the limits of merely economic treatment and involve consideration of the whole environment, cultural, social, and institutional as well as economic, in which change takes place.

---

[1] The now popular theory that a lack of balance in an economic system, the existence in it of sectors which are out of step with the others, is a stimulus to growth is argued in theoretical terms by A. O. Hirschman, *The Strategy of Economic Development* (New Haven, 1958), especially pp. 65–70.

This wider view does not necessarily conflict with the rigorous and largely quantitative analysis of the economic factors involved for which Ashton called over two decades ago and which has been in progress ever since, for other countries as well as Britain.[1] Width and imagination seem likely to become more characteristic of work in this field than they have been in the past. More and more attention is concentrated on processes of change, themselves the outcome of complex interrelations between different factors, rather than on institutions, even purely economic ones. The emergence of an important body of dynamic theory dealing with economic development has greatly aided this process. More and more light, in all probablity, will be thrown on eighteenth-century Britain by considering her in the context of a Europe much of which was also increasing its trade, urban growth, and handicraft manufactures, developing its agricultural production, and improving its communications.[2] The methods of the highly quantitative economic history, or 'cliometrics', recently developed in one or two American universities have not so far been applied in any significant way to the early stages of the Industrial Revolution. In the one limited area of eighteenth-century economic history where they have been used, that of calculating as exactly as possible the burdens imposed by the Navigation System on the American colonies before independence, they do not seem so far to have contributed much to knowledge.[3] Whether another aspect of the 'new' economic history, its use of 'figments' and 'counterfactual' situations, may eventually contribute more remains to be seen.[4] But no branch of present-day writing on the eighteenth century, with the possible exception of the all-embracing French regional studies,[5] offers greater scope than economic history, and perhaps particularly the study of the growth of industry, for the combination of width of view with rigour in analysis.

[1] T. S. Ashton, 'The Treatment of Capitalism by Historians', in F. A. Hayek (ed.), *Capitalism and the Historians* (Chicago, 1954), pp. 57–8, 60. For a recent discussion of French work of this type see T. J. Markovitch, 'La Révolution industrielle: le cas de la France', *Revue d'histoire économique et sociale*, vol. 52, No. 1 (1974), 115–25.

[2] Rostow, 'The Beginnings of Modern Growth', 573–4.

[3] C. M. Walton, 'The New Economic History and the Burdens of the Navigation Acts', *Economic History Review*, 2nd ser., xxiv, No. 4 (1971), 533–42.

[4] For a discussion of these new approaches see F. Redlich, ' "New" and Traditional Approaches to Economic History and their Interdependence', *Journal of Economic History*, xxv, No. 4 (1965), 480–95.

[5] See above, pp. 51–2.

KING, PARLIAMENT, AND PARTIES: THE PROBLEM OF THE
FIRST YEARS OF GEORGE III

The politics of the early years of the reign of George III, of the
decade 1760–70 which ended with the establishment of Lord North
as chief minister, have been more intensively discussed by historians
than those of any comparable period of British history. The difficulties and failures which these years witnessed, the party struggles,
governmental instability, and decline in the quality of political leadership which they saw, drew the attention of many writers in the nineteenth century and have, in recent decades, provided subject-matter
for some of the most meticulous of historians. 'A corpus of research',
one of the latter has claimed, 'has been built up for the reign such as
exists for no other period of British history';[1] and the assertion is
undoubtedly justified. Moreover the production of this imposing
mass of detailed scholarship, whose thoroughness and minuteness
sometimes verges upon the tedious, was preceded a century earlier
by the publication from the 1840s onwards of a remarkably extensive
body of printed primary materials relating to the political life of the
period. 'Few things are more richly illuminated by printed documents
than the politics of George III's reign', noted another modern historian;[2] and this claim is equally justified. Yet until within the last few
decades this imposing body of raw materials, and still more much
of the writing based upon it, were used largely to prolong the
currency of traditional misunderstandings and prejudices. Until
then most of the writing about these years illustrated all too clearly
how durable a myth about the past can be, once established, and how
difficult it is to shake its hold on men's minds.

The traditional accusation against the king, which reached its full
development only from the mid-nineteenth century onwards, was that
he was from the moment of his accession to the throne in 1760 determined, largely as a result of the influence of his German mother and
of his teachers in youth, to alter radically the structure of government
as it had existed under the first two Georges. The power of the crown
was to become once more a real and indeed dominant factor in the
political picture. This was to be achieved by the king's assertion of his
right, forfeited in practice under George I and George II, freely to
choose his own ministers unhampered by any consideration of the

[1] J. Brooke, *King George III* (London, 1972), Introduction, p. xvi.
[2] R. Pares, *King George III and the Politicians* (Oxford, 1953), p. 1.

party balance in the House of Commons. These ministers were to be maintained in power by the creation in Parliament, by the distribution of places, pensions, and other favours, of a body of supporters who could be relied on to back unquestioningly royal policies directed to breaking down constitutional restrictions on royal power. This view has deep historical roots. As early as 1770, in his *Thoughts on the Cause of the Present Discontents*, Edmund Burke spoke of 'a new project . . . devised by a certain set of intriguing men, totally different from the system of administration which had prevailed since the accession of the House of Brunswick';[1] perhaps the earliest allegation, by a writer of real importance, of an effort by the king and his supporters to subvert the constitution. Nevertheless this type of attack on George III was not to become a dominant theme in historical writing for the best part of a century to come. During his own lifetime much, indeed most, of the comment on him by serious writers was favourable, sometimes strongly so.

The historian who most strikingly and effectively embodied this attitude, and who was to set the tone for conservative writing in defence of the king throughout much of the nineteenth century, was John Adolphus. In his *History of England from the Accession of George the Third to the Conclusion of Peace in the Year One Thousand Seven Hundred and Eighty Three* (London, 1802) he made clear his support for George and his policies. The king, he claimed,

has tempered a noble desire to preserve from degradation the authority he inherits, with a firm and just regard to the constitution and liberties which conducted him to the throne. . . . Far from thinking that the aims of successive administrations have been directed to overthrow the liberties and constitution of the country; I am persuaded that liberty has been better understood, and more effectively and practically promoted during this period, than in any which preceded.

Because they were foreigners by birth and education, and preoccupied by the needs of the electorate of Hanover, and because they were faced by a genuine Jacobite threat, George I and George II had been forced, Adolphus argued, into dependence on a small group of great noble families who came to dominate political life and to a large extent the throne itself. George III, however, 'exempt from foreign partialities' and freed from the Jacobite danger, was able to throw off this humiliating dependence. His minister and favourite, the Earl of Bute, whose influence during his short period of office in 1761–2

[1] *Works*, ii (1803), pp. 231–3, 256.

aroused criticism and hostility of extraordinary violence, had tried merely to restore the independence of the crown 'by a moderate exertion of the constitutional prerogative'. Though Bute, because of his personal characteristics and Scottish birth, was not the right man to carry out such a policy 'the plan itself was well conceived and necessary'.[1] Here we have what was to remain the central argument of almost every later defender of the king for over a century—that he had attempted, however clumsily, merely to exert in practice powers and rights with which the constitution undoubtedly endowed him, that he had justifiably tried to escape from the clutches of the noble oligarchy which had threatened under his grandfather and great-grandfather to reduce the monarch to a mere figurehead. This had been a 'firm established phalanx, which, while it supported, obscured the throne'.[2] Unaffected by continental ideas of despotism, purely and patriotically British in outlook, George had wished to do no more than reassert the principles of the constitution against those who had in the past manipulated it for their own ends. He had 'sought, by abolishing party and national distinctions, to reign, indeed, king and protector of all his people'.[3]

In so far as the king's efforts had failed, Adolphus argued, it was not because of any inherent defects but because of the selfish irresponsibility of much of the opposition to him. London in the 1760s had shown a 'factious and overbearing spirit of resistance to the exertions of government', while 'the licentiousness of the press . . . became unbounded, and disgraceful to the nation'. This party spirit had led to deeply unfair judgements of George and his policies, for 'such were the effects of a constant and acrimonious opposition, that not only the prudence of his measures, but the purity of his intentions, was doubted'.[4]

All these claims of Adolphus were echoed by other writers in the following half-century. Thus Robert Bissett, writing at almost the same moment, also defended Bute and asserted that 'Candour must allow, that the comprehensive principle on which his majesty resolved to govern , was liberal and meritorious, though patriotism may regret that he was not more fortunate in his first choice [i.e. of minister].'[5]

[1] i. Preface, ix, 14–15.
[2] i. 362.
[3] i. 360.
[4] i. 127, 131, 360–1.
[5] R. Bissett, *The History of the Reign of George III, to the Termination of the Late War* (London, 1803), i. 362, 364.

In the 1840s another writer, Revd. T. S. Hughes, strongly denied one of the most serious of the charges levelled against the king by Burke—that he had had a body of personal advisers and favourites outside the cabinet who had wielded a real power which his ministers were denied. 'Few . . . at present are found', Hughes concluded, 'who believe in that interior cabinet behind the throne, over which a mysterious power was supposed to preside, counteracting the plans of nominal ministers, and paralysing the efforts of struggling patriotism.' George had, admittedly, been obstinate in defence of the royal prerogative; but 'if we contemplate the perilous times in which he lived, and compare his upright character, with the meanness, insincerity, and profligacy of contemporary sovereigns, we may see good cause for honouring the memory of George III'.[1] Almost at the same time Lord Mahon, in the most ambitious effort at a history of eighteenth-century England to appear until the 1870s, adopted completely the view, given its first definitive statement by Adolphus, that the king had been driven in the national interest to break the power usurped by a selfish aristocratic oligarchy. At the beginning of his reign

A small knot of grasping families among the Peers—which wished to be thought exclusively the friends of the Hanover succession, and which had hitherto looked upon Court offices, honours, and emoluments as almost an heirloom belonging to themselves—viewed with envious eyes the admission of new claimants, not as involving any principle of politics, but only as contracting their own chances of appointment.

From any comparison with this selfish crew the 'King's Friends', the body of parliamentary supporters which George built up in the early and middle 1760s, emerged very favourably. They were inspired not by the desire for offices or pensions, as Burke had unfairly alleged, but by genuine loyalty to the crown. Lord North's government in 1770–82, when George had overcome or at least greatly weakened the political factions which had given him so much trouble in the 1760s, 'freely welcomed to its high places high ability however unconnected'; whereas the Rockingham Whigs, the aggressively high-minded political faction whose propagandist Burke had been, 'almost avowedly regarded power as an heirloom in certain houses'.[2]

[1] Revd. T. S. Hughes, *The History of England, from the Accession of George III, 1760, to the Accession of Queen Victoria, 1837* (3rd edn., London, 1846), i. 177, 206.

[2] *History of England from the Peace of Utrecht to the Peace of Versailles* (5th edn. London, 1858), iv. 214–15, v. 120, vii. 143.

There was no doubt in the minds of these writers, none the less, that the 1760s had been an age of marked political corruption. In particular it was agreed that the general election of 1768 had seen the buying and selling of seats in the House of Commons, fuelled by the lavish use of some of the great fortunes now being amassed in India, carried to unprecedented and scandalous lengths. '*The current price of Boroughs*', wrote an annalist before the end of the eighteenth century, 'was enormously raised by the rival-plunderers of the East and the West, who, by a new species of alchymy, had transmuted into English gold *the BLOOD of AFRICA and the TEARS of HINDOSTAN*. Many private fortunes were ruined, or materially impaired, by contests carried on with the utmost shamelessness of political depravity.' The same judgement, in rather less emotive language, was passed by more serious historians.[1] Not all writers of the late eighteenth and early nineteenth centuries, moreover, shared the balanced and relatively favourable view of Bute taken by Adolphus and Bissett. Some remnant of the hysterical popular hostility which he had encountered in office perhaps appears in Belsham's denunciation of him as 'a nobleman haughty in his manners, contracted in his capacity, despotic in his sentiments, and mysterious in his conduct'.[2] Nevertheless the prevailing climate of opinion until the mid-nineteenth century was in general approving of what the king had tried, not without success, to achieve in the early years of his reign. In part these favourable judgements reflected George's great personal popularity during his lifetime. In part also they were the result of the long period of Tory ascendancy in British politics during the half-century before 1830. To a party to which, especially in the aftermath of the French Revolution, the phrase 'Church and King' had real meaning and emotional weight, George's efforts to assert constitutional rights which he undoubtedly possessed, together with the piety of the 'good old king' and the purity of his personal life, were very attractive. Above all, however, these writers were still fairly close in time to the 1760s. They lived in a political

[1] W. Belsham, *Memoirs of the Reign of George III to the Session of Parliament ending A.D. 1793* (London, 1795), i. 232: cf. Adolphus, *History of England*, i. 363; Mahon, *History of England*, v. 190–1; E. Holt, *The Public and Domestic Life of His Late Most Gracious Majesty, George the Third* (London, 1820), i. 159. The last of these is, however, a trivial and anecdotal work.

[2] *Memoirs of the Reign of George III*, i. 7. For a modern discussion of Bute's extreme unpopularity and the degree of justification for it see J. Brewer, 'The Misfortunes of Lord Bute: A Case Study in Eighteenth-Century Political Argument and Public Opinion', *Historical Journal*, xvi (1973), 3–43.

environment similar enough to that of these years (even Mahon began writing his *History* before the parliamentary reform of 1832) to judge the king by the standards of his own day and not by those of a later and very different age. To them the idea that practical politics might, and perhaps must, include some element of corruption, and that the crown was entitled, indeed obliged, to play an active political role, was neither strange nor unacceptable.

Even Whig writers down to the 1850s, however much they might attack George as a person, usually agreed that his actions in the early decades of his reign were not unreasonable in terms of his own ideas and the assumptions of his age. Lord Brougham, himself one of the most vocal and many-sided of Whig poiticians, had no doubt that 'In all that related to his kingly office he was the slave of deep-rooted selfishness; and no feeling of a kindly nature was allowed access to his bosom, whenever his power was concerned, either in its maintenance, or in the manner of exercising it.' Yet he had also no doubt that the king 'only discharged the duty of his station by thinking for himself, acting according to his conscientious opinion, and using his influence for giving these opinions effect'.[1] So great a Whig intellectual as Macaulay, though violently hostile to the 'King's Friends'—'a reptile species of politicians'—drew of George and his ambitions in his *Essay on the Earl of Chatham*, first published in 1844, a picture a good deal more balanced than might have been expected. In the same way a decade later, William Massey criticized the king as a man even more severely than Brougham had done but agreed, in terms which Adolphus would have found perfectly acceptable, that 'If he resorted to mystery and secret influence, it was not for the purpose of setting up a cabinet within a cabinet: but simply to disperse the haughty cabals which had enthralled his predecessors, and to recover what he thought fairly belonged to a king—the right, namely, of choosing his own servants, and being their master, instead of a puppet in their hands.'[2]

By the 1860s, however, this position was changing. The crown was

[1] *Historical Sketches of Statesmen who flourished in the Time of George III*, 1st ser. (London, 1839), pp. 6, 14.

[2] W. Massey, *A History of England during the Reign of George the Third* (London, 1855–63), i. 61, 67. The same acceptance of the view, stemming from Adolphus, that George had been inspired above all by the desire to break down the dominance of a few great Whig families, established during the two preceding reigns, can be found in a concise and pointed form in T. Erskine May, *The Constitutional History of England since the Accession of George the Third, 1760–1860* (London, 1861), i. 8–11.

now no longer a dominant force in political life. Royal and governmental patronage, though hardly negligible, no longer had the importance they had possessed a century earlier. Electoral corruption, though not yet a thing of the past, was under increasingly heavy and effective attack. A whole new series of constitutional conventions had developed and 'the spirit of the constitution' had changed in very important ways. There was thus a strong temptation for writers to judge the king, not in terms of his own day and its standards, but in those of mid- or late-Victorian England, where very different assumptions held sway. From the effects of this subtle pressure towards essentially unhistorical judgements few of them escaped.

Nevertheless the first significant and sustained attack on Adolphus and the generally favourable views of George which he had fathered was little influenced by these political factors and was based on truly historical arguments. It was the work of William Smyth, Professor of Modern History at Cambridge, who in his *Lectures on Modern History* made the first serious attack on the argument that the king had good reason to fear dominance of the throne by the great Whig families. George I and George II, he claimed, had not been in the position of humiliating dependence on the Whig leaders which Adolphus had postulated. The monarchy and the Whig lords had rather been interdependent, for 'Did not the Whig ministers and their sovereigns think the power and prosperity of each necessary to the best interests of the other?' Sir Robert Walpole's greatest fault, for example, he argued with much point, had not been a tendency to overshadow the throne but rather a 'too great anxiety for the personal favour of his sovereign; a too great readiness to make sacrifices to obtain it; an almost puerile terror of losing his place'. In fact the king had been genuinely thwarted by the politicians on only two occasions before the 1760s—in 1744, when the Pelhams drove Carteret from office against the will of George II; and in 1757, when Pitt had to be given office. The Whig leaders had never attacked the royal prerogative or shown themselves disrespectful to the ruler; George III and Bute were therefore not restoring genuine powers of the crown which needed to be defended but positively attacking the spirit of the constitution. The result of their policies, had they been completely carried out, would 'ultimately be the appearance in our own government of that temper and general servility which mark a government more or less arbitrary like the old government in France under Louis XIV'. Their success would have meant 'the extinction of every thing rare and precious in the constitution of our government'.[1]

In his view of the relationship between the throne and the political factions under George I and George II Smyth showed real perception; his picture of the situation is much more in agreement with the facts as seen by modern historians than that drawn by Adolphus and his followers. But the attacks on George III which were growing in number and severity in the later nineteenth century were based for the most part not on this type of argument but on political and moralistic considerations. In them any effort at an exact historical estimate of the balance of power between crown and parties at any time in the eighteenth century plays very much second fiddle to condemnations of the king's motives and above all of his methods. These attacks, and the absence for long of any effective rebuttal of them, meant that the Whig interpretation of the events of the 1760s was now increasingly in the ascendant.

The most violent of these later nineteenth-century critics asserted roundly that 'The object of George the Third was to make his will as absolute in England as that of any German prince over the boors and servile nobles in his dominions', and that 'The scheme of establishing his power on the ruins of our constitution . . . George the Third steadily and systematically acted upon.'[2] This was an excess of condemnation which few other writers attempted to equal. Most of them were still willing to admit that the king had many good personal qualities, even though the effects of these had been partly nullified by an inadequate education and the malign influence of his mother and Bute. As Lecky put it, 'Unlike his two predecessors, he was emphatically a gentleman.'[3] Moreover there was still some willingness on the part of George's critics to admit that the earlier dominance of the crown by the power of the Whig factions partly excused his anxiety to reassert its powers; and Lecky at least was balanced enough to admit that his methods had not been unconstitutional in any strict

---

[1] *Lectures on Modern History* (Cambridge–London, 1840), ii. 331–41.

[2] J. G. Phillimore, *History of England during the Reign of George the Third* (London, 1863), i. 273, 283.

[3] W. E. H. Lecky, *A History of England in the Eighteenth Century* (new edn., London, 1892), iii. 169–71; cf. the generally favourable view of George's character in the otherwise fiercely critical G. O. Trevelyan, *The Early History of Charles James Fox* (London, 1880), pp. 95–8; and in W. Hunt, *The History of England from the Accession of George III to the Close of Pitt's first Administration (1760–1801)* (London, 1905), pp. 3–5.

sense.[1] But the bulk of this body of writing was based on the assumption that George had on his accession inherited a system of government by responsible ministers working through parliament, a system which was recognizably the ancestor of the Victorian constitution, and that this he had attempted to destroy in the interests of his own personal power. Such an attempt could only be wholeheartedly condemned. Judgements of this kind rested upon a view of the constitution as it had existed in 1760 which was in many ways anachronistic and mistaken. But by the later decades of Victoria's reign England was far enough removed in time from the early years of George III for the anachronism to seem plausible even to intelligent men. At the very best, even relieved from absolute moral odium, George's effort to reassert his prerogative and make his constitutional powers a political reality had been, it was argued, deeply unrealistic, and therefore to be condemned. It had been an effort to stem, indeed reverse, a tide of constitutional change which was tending to give greater power to Parliament and eventually to the enlarged electorate of late nineteenth-century England, and which could not be resisted. The king had been fighting against history, and also against at least the spirit if not the letter of the constitution. He had been struggling against the forces of political progress which were marching towards the supremacy of the House of Commons and the increasingly unpolitical monarchy to be seen in late Victorian times; and whatever temporary and tactical victories he might win, such a battle he was certain in the end to lose. Moreover he had fought it by the systematic use of corruption on a scale scarcely surpassed in the history of British political life. By bribes, pensions, titles, favours of all kinds, by the unscrupulous use of crown and court patronage, he had built up in the House of Commons a group of essentially mercenary supporters, and in the process degraded still further the already low standards of eighteenth-century political life.

It is perhaps most of all in this type of criticism that the failure of most writers of the later nineteenth and early twentieth centuries to understand the political realities of the 1760s emerges clearly. They lived in an age in which the extension of the electorate, voting by ballot, and perhaps the progress of education were (though more slowly than many of them liked to admit) eliminating corruption and influence of the old kind from parliamentary elections and politi-

[1] Hunt, p. 610; Lecky, vii. 178.

cal life generally. They therefore found it both easy and tempting to condemn standards and practices which in the very different England of the eighteenth century had been unavoidable. They assumed much too uncritically that the eighteenth century not only should but could have been like their own age. The result is a tone of moral condemnation and superiority which had been much less obtrusive in the judgements passed by earlier writers. George had attempted, claimed one critic, 'to make corruption . . . do the work of an exploded and obsolete prerogative'. He was an artist in corrupt practices, alleged another, for 'He furnished the means, and minutely audited the expenditure, of corruption'; and 'He was at home in the darkest corners of the political workshop, and up to the elbows in those processes which a high-minded statesman sternly forbids, and which even a statesman who is not high-minded leaves to be conducted by others.'[1]

This generally and sometimes bitterly hostile view of the king and his policies was made more persuasive by the fact that it could now be based on an increasing body of easily accessible contemporary evidence. From its publication in 1845 Horace Walpole's *Memoirs of the Reign of George III* became a factor in influencing opinion against the king; from it the view of George as trying to destroy a free and liberal constitution which he had inherited drew powerful support. It was only one example, however, though from its stylistic qualities a very important one, of a large body of letters and memoirs relating to the early years of George III's reign which appeared in print with remarkable suddenness in the middle decades of the century. In 1838–40 there was published the *Chatham Correspondence;* in 1842–6 the *Bedford Correspondence*; in 1852 the *Rockingham Memoirs;* in 1852–3 the *Grenville Papers*; in 1853–7 the *Memorials of Charles James Fox*. This mass of material did not merely add to knowledge; it also tended to weight the scales of historical judgement against the king, since a very high proportion of it reflected the views of his political opponents. It is also true that in one or two instances at least its editing was biased against George; that of the *Bedford Correspondence* by Lord John Russell, a leading Whig politician, is the clearest example of this.[2]

[1] Phillimore, *History of England*, i. 565; Trevelyan, *History of Charles James Fox*, pp. 127, 130–1.

[2] H. Butterfield, *George III and the Historians* (London, 1957), pp. 97–107. This book is the only detailed survey of the evolution of historians' ideas about the early years of the reign of George III.

By the beginning of this century, then, the workings of the conventional wisdom had produced a view of George III which was hostile in political though not invariably in personal terms. But this rather simple-minded assessment was now to be refined and complicated by new writers. Increasingly its tones were to be muted, its hard edges blurred, by the sheer accumulation of new information and by the introduction into what was still a deceptively straightforward picture of doubts, qualifications, and refinements. This happened notably in 1907, when the German historian Albert von Ruville produced, in his *William Pitt, Earl of Chatham*, incomparably the most detailed and scholarly study hitherto published of any great figure of eighteenth-century England. Unlike virtually every book so far mentioned here, it was based on really extensive use of unprinted materials (the Newcastle and Chatham papers in the British Museum library, now the British Library, Reference Division, and documents from the Public Record Office and the Prussian archives) as well as on all the main printed sources.[1] Apart from this impressive range of knowledge von Ruville had the great advantage of seeing English politics as a foreigner, from the standpoint of a tradition quite different from the English one. He agreed that George had tried 'to break the power of the Whig aristocracy and to put an end to the party system'; but this did not arouse in him the sense of moral outrage which it had inspired in so many earlier English writers. On the contrary: at his accession 'The king's objects were both rational and opportune and he did not mean at all to infringe the constitution.' Von Ruville was well aware, unlike many of his predecessors, that the constitutional structure of England in 1760 was very different from what it had become a century or more later. He could thus write, with perfect justification, that 'George had a highly developed sense of justice, and hence was imbued with . . . respect for the law and the constitution.'[2]

However the most important development of the early twentieth century was a slow but very significant change in the interests of historians. Aspects of the subject previously disregarded and largely unknown began to attract their attention. As in other fields, new

[1] Adolphus, whose book is a work of genuine scholarship, had used some manuscript materials, notably the diary of the minor politician, Bubb Dodington, in his discussion of the 1760s; but hardly any nineteenth-century author made a real effort in this direction.

[2] A. von Ruville, *William Pitt, Earl of Chatham* (London, 1907) ii. 314, iii. 3, 8.

writing now moved away from a preoccupation with individuals and even with the policies of governments. In one direction the study of the politics of the 1760s looked with much greater attention to the specific, the concrete, the local. The eighteenth-century House of Commons and the ways in which its members were chosen began to receive much more detailed and scholarly discussion than ever before.[1] This meant that the profound differences which separated it from its successors of the nineteenth and twentieth centuries became much clearer than in the past. That contested elections were decidedly rare in the eighteenth century, that M.P.s were not chosen because of their attitude to questions of public policy, indeed that many such questions which were later to bulk very large simply did not exist at all during this period so far as Parliament was concerned, were all facts which now began to stand out with a clarity hitherto obscured by clouds of rhetoric and customary assumptions.[2]

The transition to a distinctively modern view of eighteenth-century politics, the establishment of new and realistic criteria by which to judge them and of new scholarly standards by which to weigh discussion of them, was above all the work of Sir Lewis Namier. In his *The Structure of Politics at the Accession of George III* (London, 1929) and its sequel, *England in the Age of the American Revolution* (London, 1930), he launched against the Whig interpretation of the events of the 1760s an attack which destroyed it as a view tenable by serious historians. The attack was not completely unprecedented. There had been premonitions of it in the work of some earlier writers. But Namier, armed with a knowledge of the parliaments and personalities of the first years of George III far more detailed than any possessed by his predecessors, and equipped with a mordantly brilliant style, was able to undermine traditional ideas and assumptions far more completely than anyone who had gone before him. Unlike most earlier commentators on the subject, he attempted no sustained narrative of the events of the period. Instead he produced in two remarkable books what was, in effect, a series of separate though inter-connected essays on the membership and working, in the early 1760s, of the House of Commons, which he rightly regarded

---

[1] E. and Annie G. Porritt, *The Unreformed House of Commons* (Cambridge, 1903) assembled a great mass of information of this kind and can be regarded as the first serious study of the subject.

[2] See W. T. Laprade (ed.), *The Parliamentary Papers of John Robinson, 1774-1784* (London, 1922), Introduction, pp. viii–xi for one of the first important statements of these points.

as a microcosm of the ideas and aspirations of the British 'political nation'.

Much of Namier's achievement was negative. He is important above all because he swept away myths or half-truths which had long been current and which earlier historians had failed to scrutinize closely enough. The picture of British political life which he drew was one from which coherent parties, which nineteenth-century writers had translated backwards from their own age into the 1760s, were sternly excluded. 'There were no proper party organizations about 1760', he wrote, 'though party names and cant were current; the names and the cant have since supplied the materials for an imaginary superstructure.' This assertion he supported by a careful study of the parliamentary representation during these years of Shropshire, which showed how little was meant there by the rather tattered labels 'Whig' and 'Tory'.[1] The parliaments of the period were largely mosaics of small and loosely-knit groups, sometimes in some broad sense political but often based upon merely territorial considerations or family connections. On one side of these stood the considerable number of independent country gentlemen in the House, who had no desire for power; on the other a number of supporters of the court and administration who were seldom out of office.[2] His picture was also one which found no place for political ideals, or even ideas. Men were not drawn into politics in the eighteenth century by the desire to advance a cause, to ensure the victory of a political programme, or the triumph of some distinct view of the constitution, but for essentially personal reasons. They wished above all to 'make a figure', and 'no more dreamt of a seat in the House in order to benefit humanity than a child dreams of a birthday cake that others may eat it'.[3] The Houe of Commons, to him, was to be analysed not in terms of the political outlook of its members in some abstract sense but rather in those of personal circumstances, personal connections, personal loyalties and aspirations. Some men might enter it in search of an outlet for ambitions unable otherwise to express themselves; others (notably lawyers) in quest of professional advancement; others again as professional administrators with a primary loyalty to the crown, introduced into political life that they

[1] *Structure of Politics*, i. Preface, p. vii; ii. chap. v.

[2] Namier's final thoughts on the structure of the eighteenth-century House of Commons are given in his Romanes Lecture for 1952, 'Monarchy and the Party System', which is printed in his *Personalities and Powers* (London, 1955).

[3] *Structure of Politics*, i. 4.

might serve the king or a minister more effectively. None, however, came to it as the servants of anything so tenuous and in eighteenth-century conditions so unreal as a political programme or an ideology. Namier's dominant interest in the personal, the individual, the local, was almost certainly strengthened by the influence of Freudian psychology, which was becoming increasingly important in intellectual circles in Britain in the 1920s and 1930s. This interest produced some of the most brilliantly penetrating sketches ever written of the characters and mentality of leading (and sometimes minor) figures in British politics.[1]

The bulk of his two major works, however, was devoted to a radical recasting of accepted ideas about the electoral structure of England in the 1760s. Both the influence of the king and the administration in this respect, so frequently and vitriolically attacked, and that of the great Whig magnates, were shown by detailed study and scrupulous analysis to have been greatly exaggerated by generations of uncritical historians. The administration as a whole at the beginning of the reign of George III could control elections in only about thirty seats. The Duke of Newcastle, the Whig grandee and minister who had seemed to so many nineteenth-century historians the arch-exponent of electoral influence and corruption, controlled or influenced merely a dozen.[2] Secret service funds, allegedly so lavishly distributed to build up parliamentary backing for the king and his ministers, were in fact very modest by comparison with the sums often spent by individual candidates.[3]

Finally George II had not been, as Adolphus and the tradition deriving from him had argued, the prisoner of an oligarchy of magnates and the factions which they led. On the contrary he had played an active role in political management and genuinely chosen his own ministers. George III had therefore merely followed in essentials the example set by his predecessor; he had attempted nothing re-

---

[1] See for example the famous sketch of the Duke of Newcastle in *England in the Age of the American Revolution*, pp. 75–94; and for a later appreciation of a less well-known personality, Sir Lewis Namier and J. Brooke, *Charles Townshend* (London, 1964).

[2] *Structure of Politics*, i. 13.

[3] Namier's careful and detailed discussion of this point (*Structure of Politics*, i. Chap. iv) contrasts revealingly with, for example, Lecky's vague assertion (totally unsupported by precise information of any kind) that 'Crown and Court patronage was extravagantly redundant' in the 1760s and that 'enormous corruption' was employed to force the House of Commons to accept the terms of the Peace of Paris of 1763 (*History of England*, iii. 180, 368).

voluntionary but had worked within the bounds of the constitution in trying, in effect, to replace Newcastle as a party manager.

The rigour of Namier's methods extorted admiration from his most severe critics. Nevertheless the radicalism, even if it were sometimes a little more apparent than real, of much of his interpretation of the subject could not but arouse some antagonism. The most far-reaching of the charges levied against him was that he had 'taken the mind out of history' by his almost contemptuous rejection of general ideas as a determinant of men's behaviour, not merely in England of the 1760s but at all times and in all places. He had indeed explicitly argued that there is no correlation between men's motives and the results of their actions; any such relationship is a more or less fraudulent one created *ex post facto* by intellectuals to make such actions appear more reasonable than is really the case.[1] 'In analyzing individual behaviour,' wrote one commentator, 'Sir Lewis . . . was a frank irrationalist in the respectable tradition from Hume to Freud and Pareto'; while his closest follower has praised him for being 'among the first to take into history the post-Freudian conception of the mind'.[2] The claims are justified. It was in Namier's rejection not merely of Whig historiography but of Whig and liberal ideas of political motivation that his deepest originality resided. Nevertheless the brusqueness of the rejection has exposed him to accusations of caring 'too little about those higher political considerations which . . . help to turn the study of history into a political education,' and even the disciple quoted above has admitted that 'The result of Namier's work (he was aware of it himself) was to depreciate the value of history.'[3]

In spite of the attention they attracted from the moment of their appearance, for a quarter of a century Namier's two great works inspired no imitation of any magnitude. Then, however, the middle and later 1950s saw the publication of a series of important books which used the methods which he had applied to the early 1760s as a tool with which to analyse British politics in other periods of the eighteenth century. In all this work the Namierian emphasis on the study of a large number of individuals, usually unimportant in

[1] *England in the Age of the American Revolution*, pp. 147–9.

[2] J. M. Price, 'Party, Purpose and Pattern: Sir Lewis Namier and his Critics', *Journal of British Studies*, i. (1961–2), 90; J. Brooke, 'Namier and Namierism', *History and Theory*, iii. (1963–4), 339.

[3] Butterfield, *George III and the Historians*, p. 205; Brooke, 'Namier and Namierism', 345.

themselves, and the building up on this basis of a composite picture of groups and institutions, was fundamental. 'This volume', wrote one of the authors concerned of his book, 'is a study of the origin of parties, and deals mainly with the men who led and composed them.'[1] Before the Namierian revolution such a book would certainly have been concerned largely with the differing political and constitutional ideas which, the writer would have assumed, were the main, or at least an important, basis of party distinctions. Another author, writing of the 1740s, claimed that his work 'is based above all on the biographies of the 686 individuals who sat in the Commons during the life of the Parliament which assembed on 1 December 1741 and was dissolved on 17 June 1747'.[2] Sometimes this analysis of institutions in terms merely of their membership has threatened to become obsessive. In extreme cases it has inspired quantitative and statistical studies of a kind which Namier himself had hardly imagined and probably would have hesitated to approve.[3] His interest in the concrete and the local is also reflected in the attention paid by many of his followers to the analysis of parliamentary elections. Namier's detailed discussion of that of 1761[4] has been followed by others of a similar kind, sometimes even more learned and minute, of those a little later in the century.[5] Most impressive of all has been the flowering of scholarship in the eighteenth-century sections, the only ones to appear hitherto, of the great *History of Parliament* which Namier inspired.[6] These elaborate and highly-detailed works, the products of prolonged labour by teams of researchers, are essentially collections of biographies of members of the House of Commons during the period concerned, coupled with detailed analyses of these members

---

[1] J. Brooke, *The Chatham Administration* (London, 1956), Introduction, p. xiv.

[2] J. B. Owen, *The Rise of the Pelhams* (London, 1957), Preface, p. ix.

[3] e.g. G. P. Judd IV, *Members of Parliament, 1730–1832* (New Haven, 1955), which uses mechanical methods to handle an accumulation of detailed information about the more than 5.000 men who form its subject-matter. It includes elaborate tabulations of the duration of membership, economic status, education, etc. of this mass of individuals.

[4] *Structure of Politics*, i. chap. iii.

[5] For example the accounts of the 1768 election in Brooke, *Chatham Administration*, chap. ix; on a smaller scale of that of 1774 in B. Donoughue, *British Politics and the American Revolution*, (London, 1964), chap. viii; and of that of 1780 in I. R. Christie, *The End of North's Ministry, 1780–1782* (London, 1958), Part I, chap. ii.

[6] Sir Lewis Namier and J. Brooke, *The House of Commons, 1754–1790* (3 vols., London, 1964); R. Sedgwick *The House of Commons, 1715–1754* (2 vols., London, 1970).

in terms of age, social class, education, religious affiliation, etc. Their existence means that in many ways more is now known about the House during the eighteenth century than during any other period in its history.

The last generation has thus brought great gains. It has given us more minute knowledge of eighteenth-century political life and a firmer grasp of its realities. There are nevertheless signs that a full-blooded Namierite attitude is now less widespread among historians than was the case in the 1950s and 1960s. Society and politics in the early decades of the reign of George III were still altering so slowly that the sense of change over time, the essence of history, is difficult to sustain when they are studied as they have been in the last generation. What Namier and his followers have given us is a series of sometimes brilliant and always clearly defined snapshots, not a moving picture; and each of these snapshots tends to look in essentials very like the others. Over long periods the make-up of the House of Commons changed little in terms of the social origins and outlooks of its members; the result of an election was in this respect normally to leave things very much as they had been before.[1] The increasing completeness of the knowledge of many aspects of eighteenth-century politics in Great Britain which has been achieved now tends to in-hibit further effort along strictly Namierite lines. Here, more than in most other aspects of the history of the age, new research will almost certainly confirm and expand what is already known rather than yield striking new discoveries; a law of diminishing intellectual returns is clearly at work. It is noteworthy that the most recent writing on the politics of the 1760s often shows a tendency to return to traditional narrative models and to abandon markedly Namierite types of analysis.[2]

The work of the last forty years has thus made much of the nine-teenth-century argument over the meaning of the first years of George III's reign seem wrongheaded and irrelevant. Three things are clear. Firstly, the Whig view of the king as deliberately undermin-ing a well-established constitutional structure has been shaken to its foundations. It is not, perhaps completely beyond hope of some re-

---

[1] See, for example, Christie's comment on the House of Commons of 1781 that 'Judged by the general characteristics of its membership, it differed but little from its predecessors since the beginning of the reign' (*The End of North's Ministry*, Introduction, p. xi).

[2] P. Langford, *The First Rockingham Administration, 1765–1766* (Oxford, 1973), is a good example.

pair. It is still possible to argue that in the early 1760s the opponents of Bute and the king had more justification in terms of contemporary political theory than Namier and his followers have been willing to admit, and that there was by then in existence a clear concept of ministerial responsibility to Parliament which was challenged by the claim that the king could be in effect his own minister.[1] But any idea of George as aiming at something approaching a continental despotism now seems fanciful in the extreme. The most recent biography of him embodies, in its very favourable view of his character, its severe criticisms of his political opponents, and its obvious sympathy with his dislike of all his chief ministers before North, an attitude which will certainly influence the judgements of historians for many years to come.[2] Secondly, and more important, the somewhat ostentatious moralizing and the failure to understand a very different age which marked much writing of the later nineteenth century now seems remarkably naïve and superficial. It has been replaced by a much clearer realization than ever before of how different government in eighteenth-century England was from what it was soon to become. We now have a more secure grasp than any mid- or late Victorian writer possessed of the central fact that 'The Government existed, in those days, not in order to legislate but in order to govern', that its business was the preservation of public order and above all the conduct of foreign policy, not the giving of legislative form to political ideas or party programmes.[3] Finally, almost all recent work of real originality on the political life of the period has been marked by that combination of detailed information and close analysis, that appreciation of the importance of the specific and immediate local situation, which underlies so much of the best writing of recent decades on so many aspects of eighteenth-century Europe.

[1] See Brewer, 'The Misfortunes of Lord Bute', pp. 36–40.
[2] Brooke, *King George III*, pp. 76–7, 90, 124, 156.
[3] Pares, *King George III and the Politicians*, p. 4.

CONCLUSION

THE CHAPTERS which make up this book do not provide a complete picture of the development of historical thinking about eighteenth-century Europe. Many aspects of the subject have been omitted or touched on only in passing. Nevertheless these chapters are concerned with the areas in which discussion and controversy have been most active and changing ideas most apparent. In what ways, then, has the thinking of scholars about this period in the history of Europe altered and developed in recent decades?

The most obvious and significant change has been a widening of scope and the opening-up of a new range of interests. This has meant the recognition of new and hitherto relatively unexploited aspects of history as worthy of, indeed demanding, scholarly attention. Until within the last half-century, even the last generation, historical writing about the eighteenth century still flowed within channels marked out by tradition. There were still certain areas of human activity which seemed the obvious and natural field for historical study, and others which attracted no attention, either because they were difficult or, more significantly, because they were scarcely recognized as existing at all. Until recent decades the most important single current in the stream of writing was one which was political in a traditional and rather narrow sense. It dealt with the history of political events (seen normally from above, from the standpoint of those who made rather than from that of those who endured them) and political institutions. Moreover this political history was, in the nineteenth and even the early twentieth centuries, for the most part written in terms of individual states, most of them nation states and thus apparently natural and even God-given institutions. The function of much of this writing was to promote the political unity, and therefore the strength, of the state: those who produced and encouraged it often avowed this quite openly. Thus Count Thun, recommending in 1853 to the Emperor Francis Joseph the creation of an Austrian Institute for Historical Research, argued that 'intense study of national history encourages patriotism, loyalty, love and attachment to the inherited dynasty'.[1] Uniting with this national–political current and in many

[1] A. Lhosky, *Geschichte des Instituts für Österreichische Geschichtsforschung.*

ways its most important tributary was another with a much longer ancestry, that of biography. This recounted the lives of rulers and other important figures and explained events (which again meant overwhelmingly political events) in terms of their characters, ideas, and actions. It was in these areas and others allied with them, notably the diplomatic and military ones, that traditional historiography was strong. Even the history of ideas was written predominantly in biographical terms, in those of great individuals rather than in those of the unfolding of particular concepts or the gradual change of climates of opinion: the enormous nineteenth-century literature on Voltaire is the clearest indication of this.

During the last half-century, and with accelerating effect during the last generation, this picture has changed rapidly, indeed dramatically. The old fields have continued to be cultivated, with increasing assiduity and effect; but the attention of historians has also overflowed into areas hitherto relatively untouched. Historical writing has become notably more analytical, more searching, less dominated by chronological and descriptive approaches of a kind essentially unchanged since the beginnings of history as a recognizable intellectual discipline. Until within the last few decades research for the most part centred on entities—states, nations, or individuals—which could be regarded, to varying degrees, as obviously given, as clearly constituting the framework of political and social life as it presented itself to the ordinary intelligent man. Now more and more the really challenging and interesting object of study is not some pre-existing datum to be taken for granted but a question, a problem, very often one which the researcher has identified for himself and thus in a sense created. Frequently these problems transcend national boundaries and involve difficult comparisons between apparently similar but in fact often very different institutions and groups in widely differing states and societies. This increasingly questioning type of history, moreover, now draws more and more heavily on theories, and also sometimes specific techniques, borrowed from other disciplines. Sociology in particular, and also anthropology, psychology, and statistics, have contributed in this way.

So far as subject-areas are concerned, the most obvious change has been a great growth in the interest aroused by social history as opposed to other and more traditional aspects of the past. The elastic term 'social history', however, needs here to be interpreted in its widest sense. It covers not merely the externals of a social structure, its institutions, class divisions, and economic implications, but also a great complex of ideas, traditions, and beliefs, a whole largely irrational and emotional substructure, upon which very much social behaviour is based. The study of these *mentalités populaires*, of the ideas of different societies or groups about death, about happiness, about sexual behaviour, about the rearing of children, is now perhaps the most fruitful and exciting growth-point in all writing about this period.[1] One big qualification has, however, to be made. Large-scale work of this kind on social behaviour and the ideas and assumptions which underlay it has so far been in the main confined to a single country. France, endowed with a magnificent plenitude of archive materials and the most highly-developed historical profession in the world, has hitherto dominated this fascinating if often vaguely defined field of study. Elsewhere such work is still in its infancy. 'We have too long neglected some of the most vital fields of human experience, as if unworthy of a professional historian's attention', writes an English scholar. 'For eighteenth-century England there is no good history of sex, none of prostitution; not even a good history of attitudes to women. Death has been ignored, and so has food. Animals, except as part of husbandry or the meat market, have no history. And children are little better served.'[2] The complaint is justified; and elsewhere in Europe outside France the position is no more satisfactory. None the less it seems likely that, within the inevitable limits set by the materials available, it is in this general direction that many of the most important future advances of knowledge will be made.

Compared to work of this kind, with the sensitivity to a great range of comparatively new problems which it involves, Marxism, at least in the sense of a systematic and relatively rigid body of doctrine, seems now to have relatively little that is new to offer. As a part of the intellectual environment in general it will certainly continue to influence, sometimes without their being aware of it, many writers on eighteenth-century Europe. Because of its influence

[1] For references to some of the outstanding works of this kind see pp. 52 ff. above.

[2] J. H. Plumb, 'The New World of Children in Eighteenth-century England', *Past and Present*, No. 67 (May 1975).

many of them will be more conscious than they otherwise would have been of the depth of the social conflicts which underlay the surface of European societies during this period. They will continue to be aware of the problems posed by a Marxist approach to history even while very often rejecting Marxist answers to them. In France and Italy in particular, among the states of western Europe, Marxism will clearly remain for some time to come the source of the major assumptions in terms of which many historians will look at the past. Undoubtedly, historical writing in Russia and eastern Europe will continue to be cast, at least officially, in a rigid Marxist mould. (Though this will not prevent the production there of substantial and sometimes original works of scholarship which pay tribute to official orthodoxy merely in a top-dressing of references to the Marxist classics.) But it now seems that the influence of Marxism has reached, and perhaps passed, its peak. The increasingly well-understood complexities of eighteenth-century Europe (greater than those of the nineteenth century since all states and societies were as yet much more heterogeneous, more marked by regional and local variations, than a hundred years later) make it more and more difficult to fit neatly into any grandiose schema of historical explanation.

Another major new line of development has been the growing acceptance of quantification. This has meant an increasing number of efforts, occasionally somewhat uncritical ones, to express historical problems and developments in a precise quantitative form. Some striving of this kind is often implicit in the study of *mentalités*, of mass feelings and mass attitudes, which has just been referred to. The analysis and weighing of phenomena of this kind, so fascinating and yet so difficult to evaluate accurately, is always likely to involve some element of quantification. This may take the form, for example, of studies of the degree and extent of literacy in a society, of its readership of books and newspapers, of church attendance, or of fluctuations in the nature and extent of charitable bequests. A shift of emphasis from a concern with leading individuals and ruling élites to one with large numbers of ordinary people means a move-ment away from the qualitative data of personal characteristics to the at least implicitly quantitative ones in terms of which alone mankind in the mass can be effectively discussed. From sociology, moreover, history has borrowed a number of quite sophisticated techniques for the surveying of social attitudes and values and for weighing, analysing, and correlating information about them. One example of this is the use by historians, though as yet on a small scale, of 'content analysis', on which there is now a considerable sociological literature. This involves the counting, in a body of documentary or literary evidence illustrative of a more or less specific problem, of allusions, which may be either direct or indirect and symbolic, to particular ideas or values in an effort to measure their importance. The practical difficulties, notably in selecting the material to be studied, and in devising methods of counting which allow adequately for the way in which the meaning of a term may change with the context in which it is used, are considerable. Content analysis also tends to be a very laborious process, though much of the mechanical labour can be avoided by the use of computer techniques. For these reasons it has not so far been very extensively used by historians; and there are few examples of its systematic application to an eighteenth-century topic.[1] It seems unlikely that it will ever be one of the more fruitful approaches to the history of the period: nevertheless it illustrates once more both the growing desire of historians to measure rather than merely to describe or recount, and the way in which this has involved them in the use of techniques borrowed from other disciplines.

How far the movement towards quantification will endow the study of history with new creative possibilities, and how far prove in the long run something of a dead end, is not easy to say. Its potentialities must always be limited by the adequacy of the data with which it works; and these must not merely be reasonably plentiful but, often more difficult to achieve, uniform enough to ensure that like really is being grouped or compared with like. In some branches of economic history these conditions can be met. Though the 'cliometrics' exemplified by a good deal of recent work on the nineteenth century has as yet relatively little parallel in writing on earlier periods, a certain amount has already been done in this direction[2] and more will surely be done in the future. The study of eighteenth-century societies, however, much more than the economic history of the period, is confused and hampered by the existence of a whole battery of terms–peasant, noble, bourgeois, merchant—which are unsatisfactory precisely because they say far too much, because they

---

[1] The best-known is perhaps R. L. Merritt, *Symbols of American Community, 1735–1775* (New Haven, 1966), which uses content analysis of the press during these decades in an effort to measure more precisely than ever before the growth of national consciousness in the British American colonies.

[2] See e.g. the French examples referred to on pp. 40 ff. above.

bring under one verbal umbrella a variety of often widely differing realities. The bluntness of these conceptual tools tends to make precise quantitative analysis of a social structure more difficult, a more subtle enterprise calling for finer judgement, than that of the functioning of an economic mechanism. There is also an obvious danger that the mere manipulation of quantitative information, of 'hard facts' often difficult to interpret accurately and sometimes even trivial, may too easily become a substitute for historical imagination and sympathy. Certainly no amount of quantification can ever replace the semi-instinctive identification with an individual in a specific past situation, the intuitive understanding of his feelings and thoughts, which lies at the roots of so much of the finest historical writing. Yet unquestionably a striving for greater precision, quantitatively expressed, will for the foreseeable future remain one of the growth-points of research, for the eighteenth century as for other epochs. The tools it offers may be difficult to use effectively; but their potentialities must and will be fully explored. The greatest and most securely based victories of the quantitative approach have hitherto been achieved, so far as this period is concered, in historical demography; and here as in so many other areas France, thanks again largely to the excellence of the raw materials available, has taken the lead.[1] The techniques of the demographer (family reconstruction is the obvious example) are for the most part too specialized to be directly applicable to other branches of history. But the remarkable efforts at computerized studies of both military organization and social structure in eighteenth-century France which are mentioned above[2] show the same search, in principle always admirable in spite of occasional excesses and extravagances, for the maximum attainable accuracy and precision.

A final leading characteristic of recent work on many aspects of eighteenth-century Europe is its emphasis on the identifiable and coherent local unit as, for many purposes, the essential foundation for serious studies. Precise and highly detailed demographic work has confined itself for the most part to small areas, very often a single village or parish (in large part because of the practical difficulty of handling the enormous quantities of data involved if a large unit is concerned). In England, under the influence of Namier and his followers, detailed studies of parliamentary elections in single

[1] See above, pp. 43–4.
[2] See above, pp. 44–7.

constituencies, or of the parliamentary politics of a more or less narrowly defined area over a longer period, have become a well-established type of contribution to the history of politics.[1] In France both the historic provinces and the *généralités* super-imposed on them as administrative units have provided, on a larger geographical scale, the basis for the most impressive works of local history (if the English phrase is not here too restrictive in its implications) ever written.[2] Every large-scale general history of a great European state during the eighteenth century hitherto written has been weakened by the lack of a secure foundation in local and regional studies of this sort. In the future, however, any work of this kind worth producing will be very largely a combination and synthesis of such studies. The writing of history will thus reflect, far more than in the past, a reality which, in an age of limited mental horizons and undeveloped communications, was still for very many purposes obstinately local.

Eighteenth-century Europe lacks the marvellous drive and self-confidence, the exuberant expansionism and intoxicating sense of achievement, of the middle and later nineteenth century. Nevertheless the study of its history has meaning and interest for our own age. The eighteenth century has inevitably that irreducible degree of significance which any link in the chain which stretches from the present day into the more or less remote past must possess. More precisely and importantly, however, it can still intrigue us by its mixed and disparate character, by discontinuities which give it a more ambiguous relationship to the present day than most comparable epochs. In some spheres the links between it and our own age are clearly perceptible. In some ways the man of the later twentieth century can feel himself in a real and direct sense the heir, if not of the eighteenth century as a whole, at least of the period 1750–89. The ideas and still more the tone of Rousseau, for example, the potentially explosive idea, of which he is the great evangelist, of happiness as a right of which the ordinary man is continually defrauded by the

[1] For a good study of a single election see R. J. Robson, *The Oxfordshire Election of 1754* (London, 1949). For studies of the politics of particular areas see the discussion of Buckinghamshire in L. M. Wiggin, *The Faction of Cousins: A Political Account of the Grenvilles, 1733–1763* (New Haven, Conn., 1958); of an Irish constituency in A. P. W. Malcolmson, 'Election Politics in the Borough of Antrim, 1750–1800', *Irish Historical Studies*, xvii (1970); and of Northamptonshire from the 1740s to the election of 1807 in H. C. F. Lansberry, 'A Whig Inheritance', *Bulletin of the Institute of Historical Research*, xli (1968).

[2] See p. 51 above.

stupidity and malice of the world as it is, are as seductive and powerful today as they have ever been. The industrialism which had established itself in a few areas of Britain by the 1780s, again, leads by a direct and clearly traceable line of descent to the infinitely more complex and sophisticated industrial world of today. Yet these islands of modernity, intellectual or physical, are still in the eighteenth century surrounded by an ocean of tradition and conservatism, of backward-looking localism and hostility to change, which link most parts of the Continent with the Middle Ages rather than with the twentieth or even the nineteenth centuries. A local patois which in the absence of modern communications might be almost incomprehensible outside a limited area;[1] a persisting belief in many parts of Europe in the magical or quasi-magical powers of rulers;[2] garbled forms of popular Christianity and endemic popular belief in spells, witches, omens, and miraculous cures[3]—these are only some of the evidence which can be adduced in support of this argument. 'I have been in actual contact with the primitive ages', wrote the great French scholar Ernest Renan in 1883, speaking of his childhood in the 1820s. 'The most remote past was still in existence in Brittany up to 1830. The work of the fourteenth and fifteenth centuries passed daily before the eyes of those who lived in the towns.'[4] A majority of Europeans in the eighteenth century had a similar experience. This embedding of a few elements of marked sophistication and modernity in a matrix of traditionalism and conservatism is seen most strikingly in France; mainly because of the brilliance of French intellectual life during the eighteenth century, and in part also because we are simply so much better informed in many ways about France than about any other European state in this period. In England or the Dutch Republic, where society was more of a piece, where the level of popular education was higher and the peaks of

[1] Thus the *greffier* of Riom in the Auvergne had to abandon an attempt to interrogate a young beggar from Courpière, a small town little more than twenty miles distant, because 'the speech of Courpière differs considerably from that of Riom' (A. Poitrineau, *La Vie rurale au Basse-Auvergne au XVIIIe siècle*, (*1726–1789*) (Paris, 1965), i. 117).

[2] Over 2,000 sufferers from scrofula came to receive the magically healing touch of Louis XV on the day before his coronation in 1722, while over 2,400 came to beg the help of Louis XVI on the same occasion over half a century later. The Tsar Paul, at his coronation in 1796, made perhaps the most striking assertion during the century of the priestly character of a legitimate ruler by giving himself communion.

[3] See p. 54 above.

[4] *Recollections of my Youth* (Boston, 1924), p. 76.

intellectual life were less elevated, greater homogeneity is to be seen. It is this homogeneity, in economic life, in intellectual standards, and in levels of information, which is one of the truest criteria of modernity. Towards it the present-day world incessantly and consciously strives. But throughout eighteenth-century Europe it is contrasts most of all, contrasts of a sharpness not easy to imagine today, which fascinate the historian even while they impose difficulties upon him.

INDEX